In Full View

A True and Accurate Account of Lewis and Clark's
Arrival at the Pacific Ocean, and Their Search for a Winter Camp
Along the Lower Columbia River

I landed and formed a camp on the highest Spot I could find . . .
This I could plainly see would be the extent
of our journey by water. . . . in full view of the Ocian

William Clark
November 15, 1805

In Full View

A True and Accurate Account of Lewis and Clark's Arrival at the Pacific Ocean, and Their Search for a Winter Camp Along the Lower Columbia River

Rex Ziak

For:
Judy & Brian
Enjoy my "View"
Rex Ziak
2005

Moffitt House Press
Astoria, Oregon

Moffitt House Press
P.O. Box 282
Astoria, Oregon 9710

Design and production: Kate Hawley, Milwaukee, Wisconsin.
Editorial Services: Amelia and Fred Hard, Portland, Oregon.

Quotes from *The Journals of the Lewis and Clark Expedition*, edited by Gary E. Moulton,
used with the permission of the publisher, The University of Nebraska Press: Lincoln, 1983-2001.

For photo and illustration credits, see p. 216.

Second Edition, 2005

ISBN 0-9725315-1-3
Library of Congress Control Number: 2004117426

Printed in Hong Kong

In Full View *is dedicated to*

Thomas Jefferson
(1743-1826)
His wisdom and curiosity.

and

T. Rice Holmes
(1846-1937)
The scholar who taught me how to discover history.

ACKNOWLEDGMENTS

THIS IS THE MOMENT IN THE PRODUCTION OF A BOOK WHEN THE AUTHOR can thank all the people and institutions that have made the work possible. Most authors probably find this easy; however, I am indebted to such a long list of individuals and organizations that I do not know where to begin.

Perhaps I should first note my gratitude to the editors and institutions that publish the journals, letters, and maps of noteworthy people from the past. Without your painstaking work, none of us would have any idea of the past. I am humbled by the work of Gary E. Moulton and his predecessor Reuben Gold Thwaites, both of whom have edited the entire journals of Lewis and Clark, making them available to the general public. Donald D. Jackson's collections of Lewis and Clark's letters are invaluable to any student of this history, and the many fine editors of Thomas Jefferson's correspondence have given us a great gift. You, as well as all editors of all historic documents, deserve special thanks, and so do the universities and historical societies that publish such books.

To the institutions, I cannot thank you enough. The entire staff at the Fort Clatsop National Monument has been especially helpful and informative, always quick to answer any question. The Columbia River Maritime Museum staff was patient with my requests and they, along with the Clatsop County Historical Society, provided me with essential historic photographs. The Ilwaco Heritage Museum was always ready with answers to my questions; and Washington State Parks at Cape Disappointment allowed unlimited access to Lewis and Clark campsites, which was essential during my research. I cannot thank you enough.

The Timberland Regional Library promptly searched any request and provided a place for me to spread out my pages. The Manuscript and Special Collections department at the University of Washington was generous with their photographic archives, as was Seattle's Museum of History & Industry. And I am especially indebted to the staff of the Oregon Historical Society who opened the doors to its fabulous archives of maps and photography. My association with that fine cornerstone of Northwest history has been a joy.

Special thanks goes to Chuck and Grace Bartlett for providing me with such excellent elk and deer photography and to Michael Haynes for his superb depiction of Lewis and Clark, which is destined to become a classic. My friend Sonja May accepted my challenge to create a work of art that captured a moment never before seen, and she succeeded. Penny Guest and Jim McGlinn gave me free access to their fabulous collection of artifacts and artwork, and their knowledge added essential accuracy to this book.

The individuals who have inspired and assisted me from the beginning to the end of this project deserve very special thanks. Bob first lit the fire and impressed upon me the importance of publishing my discoveries in the lower Columbia. My good friend Ray walked the shorelines with me, read the journals and remained helpful from start to finish. Insights into his native American culture and art, combined with his thorough knowledge of wildlife and forests, gave me a greater dimension of understanding into the past.

M.L. was patient, insisting that I tell a good story; Paul grew impatient, but never gave up hope. Teri listened to the same stories over and over with nothing but encouragement, and Debi remained confident that my effort would add up to something good. Vince shared oral histories of salmon and the past; I hope someday to know as much as he has forgotten. Cathy was always helpful with her wealth of botanical knowledge and willingness to assist in any way; Michelle was always ready to research the most obscure of facts. Jeffrey made sure every color photograph was handled correctly and with care; Karen let us turn her folks' cabin into a production studio, and Nani gave us a place for the final assembly. To all these friends I would like to say "Thank you" for being there and for helping me in so many different ways.

The final draft of my book was read by a distinguished group. Special thanks goes to the staff of Fort Clatsop National Monument for their comments and help. Barbara, Hobe, and Bob agreed to read the text. The advice they provided was invaluable.

To my proofreader Mary, thank you for your eagle eye. Any errors which may remain are, of course, my sole responsibility.

My immediate family made tremendous sacrifices. I was so wrapped up in this project for so many years that I forgot birthdays, missed holidays, and neglected duties. My brother Larry filled in for me on countless occasions without any complaint, which freed me to concentrate on what I had to do. I owe him more than I could ever repay.

As my book approached the final months of work, it was suddenly paralyzed by two insurmountable problems. Its completion seemed doubtful. Thankfully, my good, long-time friends came to my aid. Ray stepped in to guarantee that this book would be published with the quality and the timeframe I needed. On the other front, Fred and Amelia committed to keep watch over every word. They set aside their own lives and focused full attention on making this book as good as it could possibly be. It was not their choice of subject or wording, and my last minute revisions were a challenge to keep up with; however, they committed to help and stayed with the project day after day after day. With that team behind me I was free to complete my project, and felt confident my words would express what I intended to say.

Last, but certainly not least, is my good friend and colleague Kate. She took on this project back in 1999 thinking it would be finished by 2000. Despite the additional years of research and writing, she remained interested and enthusiastic. I talked visual ideas over the phone, which she transformed into elegant layouts. I gave her an old black and white navigation chart, and she turned it into a work of art. I wrote out text and said, "I sure would like a map here and here," and she made sure maps appeared exactly where they were most needed. Without Kate's keen conceptual abilities, hard work, experience, and dedication, this book would have ended up as an incomprehensible jumble of words and photos. I would never consider taking on another project like this without her as part of my team.

I am unable to list all the colleagues, fellow board-members, and associates who have lent their resources, time, and support throughout these past years, but you know who you are, and I want to thank you.

Rex Ziak
October, 2002

Table of Contents

FOREWORD

THE MOST INSIGHTFUL WORKS ON THE LEWIS AND CLARK Expedition, indeed, all of the best histories of exploration, have come from scholars who felt the trail under their feet and the breeze against their face. Herbert Eugene Bolton cheerfully left his office at the University of California, Berkeley and doggedly retraced the journeys of missionary Eusebio Francisco Kino, S.J. across the desert southwest from Baja California to present day Tucson. Then he wrote a biography that is a classic in the field. Thor Heyerdahl stepped away from his studies in anthropology to pilot *Kon-Tiki* thousands of miles across an open ocean from Peru to the Tuamotu Islands of Polynesia following a watery route pioneered centuries earlier. Then he wrote a memoir that rivals any adventure story. Olin D. Wheeler was, perhaps, the first author of a serious work on the Lewis and Clark Expedition to revisit the trail of the Corps of Discovery, and when he did so, at the turn of the twentieth century, it was dangerous, remote and unmarked. Then he penned a two volume account of his journey that has been prominently listed on Lewis and Clark bibliographies for a century. Rex Ziak's book is considerably less than two volumes in length, and it limits its scope to only one month in Lewis and Clark's trip of a lifetime, but he otherwise shares with other great writers of exploration history an intimate knowledge of his subjects and he knows, even feels, the land they traversed.

For years Rex studied the journals of Lewis and Clark like most of us do: in a chair with a reading light over our shoulder. Then, for several more years, he walked the beaches of the Columbia River estuary imagining the canoes of Lewis and Clark keeping pace with him just off shore. He could do this often because he lives, and has always lived, right on the Columbia River, nearly within seeing distance of the great bar where the river clashes with the Pacific Ocean. Most recently, Rex has been writing the story of the Corps of Discovery. Not the whole journey, just the part he knows intimately, the final twenty-five miles of Lewis and Clark's great journey of 1804 to 1806. No one else knows as much about what took place in November and December 1805 in the Columbia River estuary as does Rex Ziak.

Numerous historians have written that Lewis and Clark's 1805 passage across the Lolo Trail in the Bitterroot Mountains of Idaho was the most dangerous of all their undertakings in a journey of 8,000 miles. No one will contest the right of any author, who has done his research, to make such a conclusive statement. But no one who reads *In Full View* will ever completely agree with that statement again. The month that Lewis and Clark spent on the shores of the Columbia River in 1805 was the most discouraging, dangerous and disagreeable period ever experienced by the captains and their corps. Tidewater washed away their camps, heavy rains and gale force winds created a life-threatening wind chill. Huge timbers tumbled ashore, nearly on top of them, and baseball-sized rocks pummeled the men from the heights adjacent to the shoreline. Rex illuminates the journals with the common sense of a life-long woodsman describing, for example, the unique attributes of "whitewater chop" and "highwater slack." For the first time, the expedition will retreat. In an unusual move, the expedition will split themselves into multiple units. Indian relationships are complex and misunderstood. Hunting is life. In a half-dozen ways the period from November 7 to December 7, 1805, is a turning point in the journey of the expedition.

Until now the jury has been out on a number of controversial matters dealing with the arrival of the Lewis and Clark Expedition at the mouth of the Columbia. Did the expedition canoes hug the shore of Gray's Bay or cut across the open water? Where is the "dismal nitch"? Where is the end of the trail, the place where Lewis and Clark satisfied the demand of President Thomas Jefferson that they intercept the Pacific Ocean? Which route did William Clark take to, and from, Cape Disappointment? What was the meaning of the "election" on November 24? Where have all the elk gone on the north side of the river? The answer to these, and other, questions are no longer a matter of conjecture. They have been asked and answered.

I sincerely hope that this work will inspire other admirers of the Lewis and Clark Expedition to do with other segments of the journey something similar to what Rex has done with the Columbia River. James Fazio preceded Rex with a study of the Lolo Trail called *Across the Snowy Ranges* (2001). But there is still a need for similar, intensive analyses of the Lemhi Trail route, the Great Falls portage, the journey of Sergeant John Ordway to the Salmon River, and many more episodes. "We proceeded on" is a common phrase in the Lewis and Clark journals, as well as a common phrase among students of the Lewis and Clark Expedition. Rex's book has helped us all to "proceed on" better informed.

Robert Carriker
Professor of History, Gonzaga University and
Director of National Endowment for Humanities seminars
on Lewis and Clark, 1985 – 2003.

PREFATORY NOTE TO THE SECOND EDITION

IN THE BRIEF TIME SINCE THE FIRST EDITION OF THIS BOOK WAS published, barely more than two years ago, there has been a growing public awareness of the western portion of Lewis and Clark's journey. Though there have been significant developments all along the trail, none is more important than the formation of the country's newest national park, the Lewis and Clark National Historical Park.

This park is the result of a major shift in our understanding of Lewis and Clark history at the mouth of the Columbia River. In the past, Fort Clatsop was seen as this area's only historic site; the names Dismal Nitch and Station Camp were completely unknown, as was the remarkable history that occurred in these places.

I am pleased to say that my research over a ten-year period brought much of this history to light. When I revealed my discoveries to the local community, my neighbors proudly claimed this lost history, and together we impressed upon state and national leaders the need to commemorate these sites. Each year more and more people became involved, until finally, after years of effort and the work of hundreds of individuals, the names "Station Camp" and "Dismal Nitch" were read aloud for the vote in the United States Congress that created the Lewis and Clark National Historical Park.

In these pages, you will find my research and the story that created America's newest national park. This history, once known only to me and my close neighbors, now belongs to all of America and so does the land on which these dramatic events occurred.

Rex Ziak
January, 2005

INTRODUCTION

I WAS BORN AND RAISED NEAR THE MOUTH OF THE COLUMBIA River and have spent almost my entire life here. The popular version of the story of Lewis and Clark's adventures in my part of the country has been familiar to me since I was a little boy. So it is easy to imagine how surprised I felt the day I realized that the full history of Lewis and Clark's arrival at the ocean had never been thoroughly researched or described.

I discovered this gap in the history of the expedition when I became interested in the final days of Lewis and Clark's westward journey. Of course, I knew that they had paddled down the Columbia River and built a winter camp near the ocean; but when I looked in histories of the expedition for the specific details of their experience here, I found nothing. The scholars had overlooked this portion of the expedition.

In order to satisfy my own curiosity, I began to carefully read Lewis and Clark's journals and to study Clark's maps. It didn't take me long to realize why this segment of their journey had not been sufficiently explained. Clark's journals are vague and confusing, and his descriptions lack specific detail. Also, Lewis wasn't writing in his journal at that point in the expedition, so scholars didn't have the benefit of his observations.

Fortunately, when Thomas Jefferson sent these men out into the wilderness, he insisted that they write down a broad range of information about the climate, the native peoples, animals, plants, and geography. Despite the lack of detail in Clark's journal entries, I found that my knowledge of this region fleshed out Clark's descriptions and gave me a special insight into Lewis and Clark's experience. In addition to this, it was easy for me to spend many, many weeks every winter walking the shoreline where they traveled, looking at the same landscape they described, and watching the rhythms and patterns of the weather. This familiarity, combined with a decade of persistence, allowed me to gain a deeper understanding of what happened to the Corps of Discovery at the mouth of the Columbia River in 1805.

Eventually, as the day-to-day drama of their ordeal began to unfold, I realized that I was seeing something unique. Compressed into a period of exactly one month were some of the most amazing adventures these men had endured. They were tossed around, robbed, starved and betrayed. They got lost in the fog, lost in the woods, lost their rifles and even lost Lewis for several days. They changed course repeatedly, buried their canoes, and narrowly escaped injury time and time again. Clark said it was the most disagreeable time he had ever experienced.

I communicated this history at first by writing essays, giving lectures and leading field trips; gradually parts of the story became known. It was never my intention to write a book about Lewis and Clark, but I eventually realized that if this portion of their story was ever to be known and appreciated by the public, then it would be up to me to write and publish it.

This book is my attempt to reveal this hidden piece of Lewis and Clark's history. I have carefully followed Clark's journal, and whenever possible let him explain what he was thinking and doing. However, the journals of Lewis and Clark and their men are like a jig-saw puzzle with more than half the pieces missing; sometimes you get the whole picture, and at other times all you have is an outline. Despite this lack of information, I have made every effort to remain as accurate as possible and to avoid speculation.

The reader will notice the extensive use of maps in this book. I drew maps as I studied this history in order to trace the complex movement of the men. I found that visualizing their movements in this way was extremely helpful for my understanding of their ordeal, and I hope the reader will also find this useful.

Concerning the Indians of the Lower Columbia River, I want to make clear that even though I grew up with and went to school with descendants of the Chinooks, I am not qualified to speak for them or to describe their existence two hundred years ago. It is impossible to recount Lewis and Clark's experiences in this region accurately without frequently mentioning the Indians they encountered, but the Chinooks and Clatsops have their own story to tell, and I certainly hope that one day they will do so.

This book is the result of trying to satisfy my own curiosity about Lewis and Clark's arrival at the Pacific Ocean. I've given it my best effort, and I hope the light it sheds on this little-known part of the expedition will satisfy the reader's curiosity, too.

Rex Ziak
October, 2002

"Direct & Practicable Water Communication"

The Northwest Passage

Thomas Jefferson was aware of the major rivers of North America; however, there were many unknowns. The Missouri River flowed into the Mississippi, but no one knew where it originated. The Columbia River had been discovered in 1792, but only the lower hundred miles had been explored.

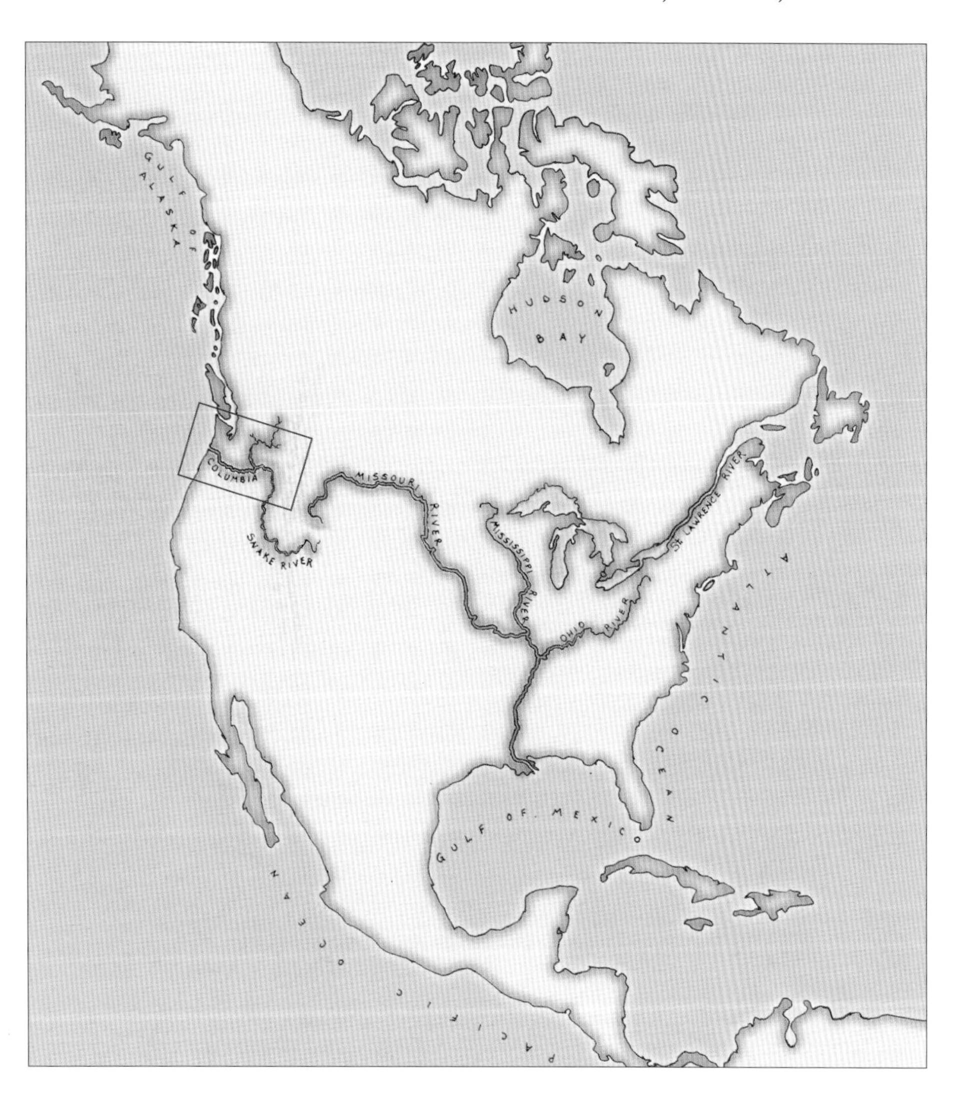

From the maps that existed, Mr. Jefferson could see that these two large rivers shared a similar latitude. Did they originate in the same watershed? If so, would it be possible to travel up the Missouri, cross over to the Columbia, and reach the Pacific Ocean?

In order to answer these questions, President Jefferson sent Meriwether Lewis to investigate. His instructions read:

> *The object of your mission is to explore the Missouri river, & such principal stream of it, as by it's course and communication with the waters of the Pacific Ocean . . . may offer the most direct & practicable water communication across this continent.*

Lewis and Clark would discover that the Missouri and the Columbia Rivers do not connect with one another, putting an end to hopes for a continuous water route across the continent. The Missouri bends to the south and divides into tributaries, whereas the Columbia originates in the Canadian Rockies. The

Columbia's major tributary, the Snake River, begins in Wyoming near the Yellowstone, flows west, then north, then west again until it joins the Columbia.

The combined Columbia and Snake River system drains a watershed that covers an area larger than France. The run-off of snow-water from the western side of the Rockies, combined with that from the Cascade Range, results in a flow of water that ranks the Columbia as North America's fourth largest river. This immense surge of water discharges into the Pacific Ocean, forming the mouth of the Columbia, "Graveyard of the Pacific."

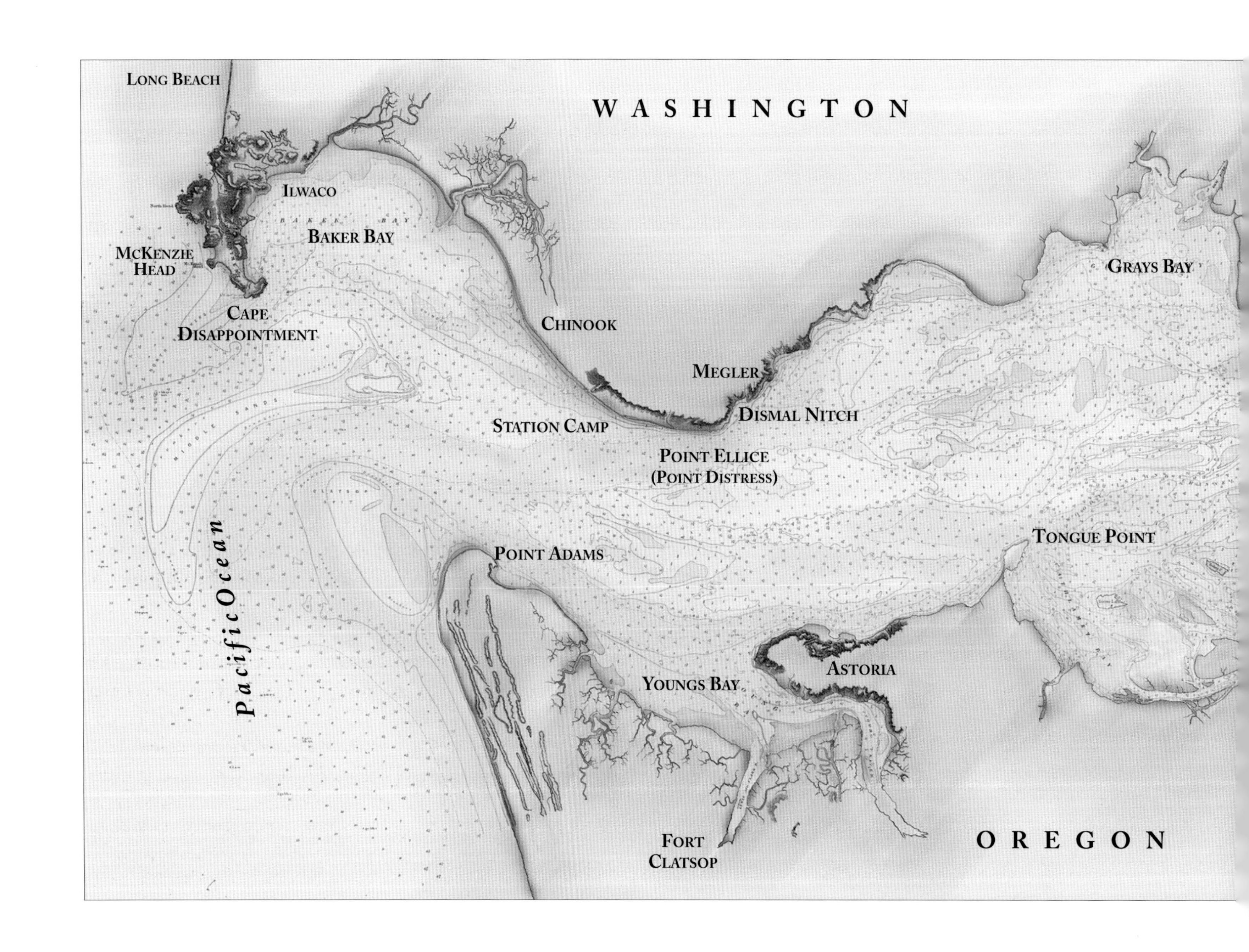
Long Beach
Washington
Ilwaco
Baker Bay
McKenzie Head
Cape Disappointment
Chinook
Grays Bay
Megler
Dismal Nitch
Station Camp
Point Ellice (Point Distress)
Pacific Ocean
Point Adams
Tongue Point
Astoria
Youngs Bay
Fort Clatsop
Oregon

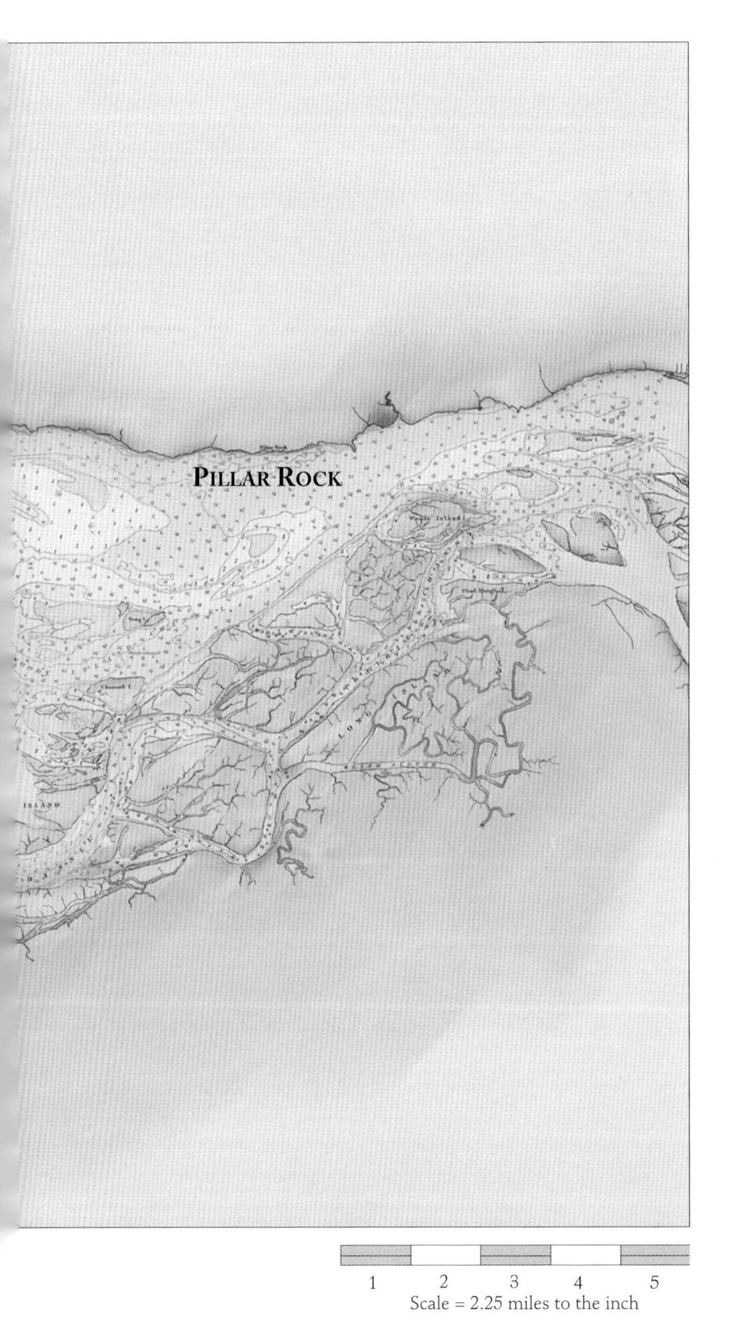

THE MAP

Navigation Chart of the Mouth of the Columbia River,
London, Published at the Admiralty, April 12, 1876.

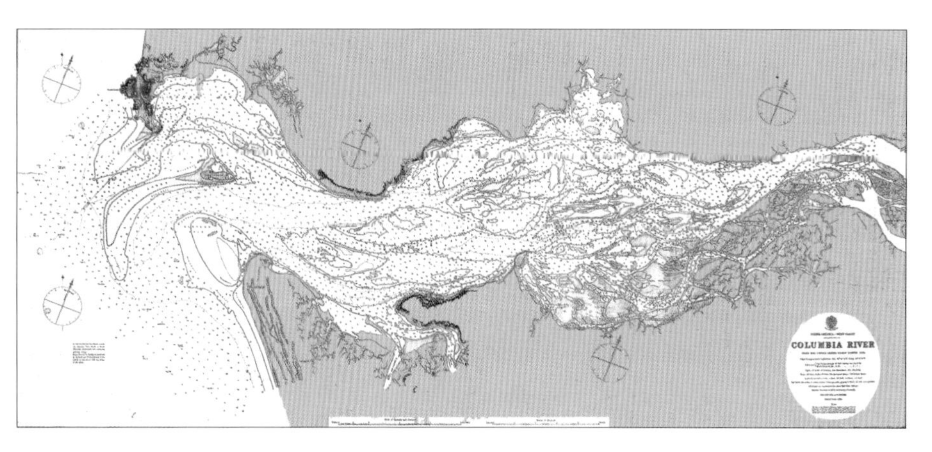

The geography of the Lower Columbia River is complex; the river is wide, and the shorelines curve around to every point of the compass. Because of this, I knew it would be essential to use a single reference map in order to explain Lewis and Clark's arrival at the ocean.

Fortunately, I came across an old navigation chart that shows the Columbia before it had been altered by jetties, dikes, bridges, and roads. This map, printed in England in 1876, shows with rare accuracy the wild lower Columbia River as it was when Lewis and Clark saw it. In respect to the contour of the shoreline, it is almost as accurate as a satellite photograph.

The original chart, shown above, was printed in black and white. In order to make it easier to read, color was added to help distinguish between shoreline and water. The names of significant sites such as Dismal Nitch, Station Camp, and Fort Clatsop, were also added to help pinpoint these locations.

PROLOGUE

THE LEWIS AND CLARK EXPEDITION BEGAN IN THE MIND OF Thomas Jefferson.

As a boy growing up in Virginia, Thomas Jefferson developed a keen interest in the plants, geography, and Indian cultures of that region. As he grew older, his knowledge and curiosity broadened. The American Philosophical Society brought him into contact with other leading intellectuals who were actively collecting, identifying, and preserving the natural wonders of North America, and from this contact sprung three early attempts to explore the West.

In 1783, Mr. Jefferson proposed an expedition from the Mississippi to California; in 1786, he suggested an exploration in the reverse direction, beginning on the west coast and traveling east. Finally, in 1793, he and several colleagues invested money to finance another western expedition; but it, like all the others, never advanced much past the planning stages.

In 1796, Jefferson was elected President of the American Philosophical Society; four years later, he was elected president of the United States. Finally empowered with the necessary influence and authority, he could fulfill his long-time ambition of exploring the western half of North America.

In 1802, the time was right to propose an exploration of the West. Those skeptical of this venture had been silenced when the Louisiana Territory was unexpectedly acquired. Suddenly, knowledge of the new land, its rivers, inhabitants, and economic opportunities became a priority. First, President Jefferson found the right man to lead the expedition: his personal secretary, Meriwether Lewis. Next, he requested $2500 from Congress to finance the venture. Then he recruited his colleagues from the American Philosophical Society to train Lewis in collecting specimens, finding latitude, and analyzing discoveries. The scope of the expedition was so extensive that Lewis invited his trusted friend William Clark to assist in these preparations and to join him on the journey itself. The two men worked together for more than a year building boats, purchasing supplies, and assembling the crews.

In May of 1804, the party, which numbered more than fifty men, loaded their equipment into three boats and set out up the Missouri River. The first summer of travel covered nearly 1600 miles before the winter weather forced them to halt and build a winter camp among the Mandan Indians.

On April 7, 1805, after the frozen river had thawed, Lewis and Clark sent part of their crew aboard their large boat back to St. Louis while they resumed their voyage up the Missouri River. Their entire party now numbered only thirty-three individuals.

Four months later, they had reached the foothills of the Rocky Mountains, where the shallow rivers forced them to abandon their canoes, purchase horses, and continue west over the mountains on foot. By late September, the party had successfully crossed the Bitterroot Mountains and found a river flowing west. Five large pine trees were carved into canoes, then launched. On October 7, they began their western river journey.

Day after day they paddled downriver braving whitewater rapids and dangerous waterfalls. Food was scarce; firewood did not exist. Their canoes flipped over and smashed into rocks, but this only delayed them until they had dried out and made the necessary repairs.

After exactly one month and 460 miles of travel downriver, the party found themselves in the broad lower Columbia River. The snow-capped mountains were behind them now, and they met Indians dressed in sailors' clothing. Gray and white gulls patrolled the river for food while low moisture-laden clouds swirled overhead. There was no sign of the ocean, but Lewis and Clark had every reason to believe that the Pacific must be very, very near. The date was November 7, 1805.

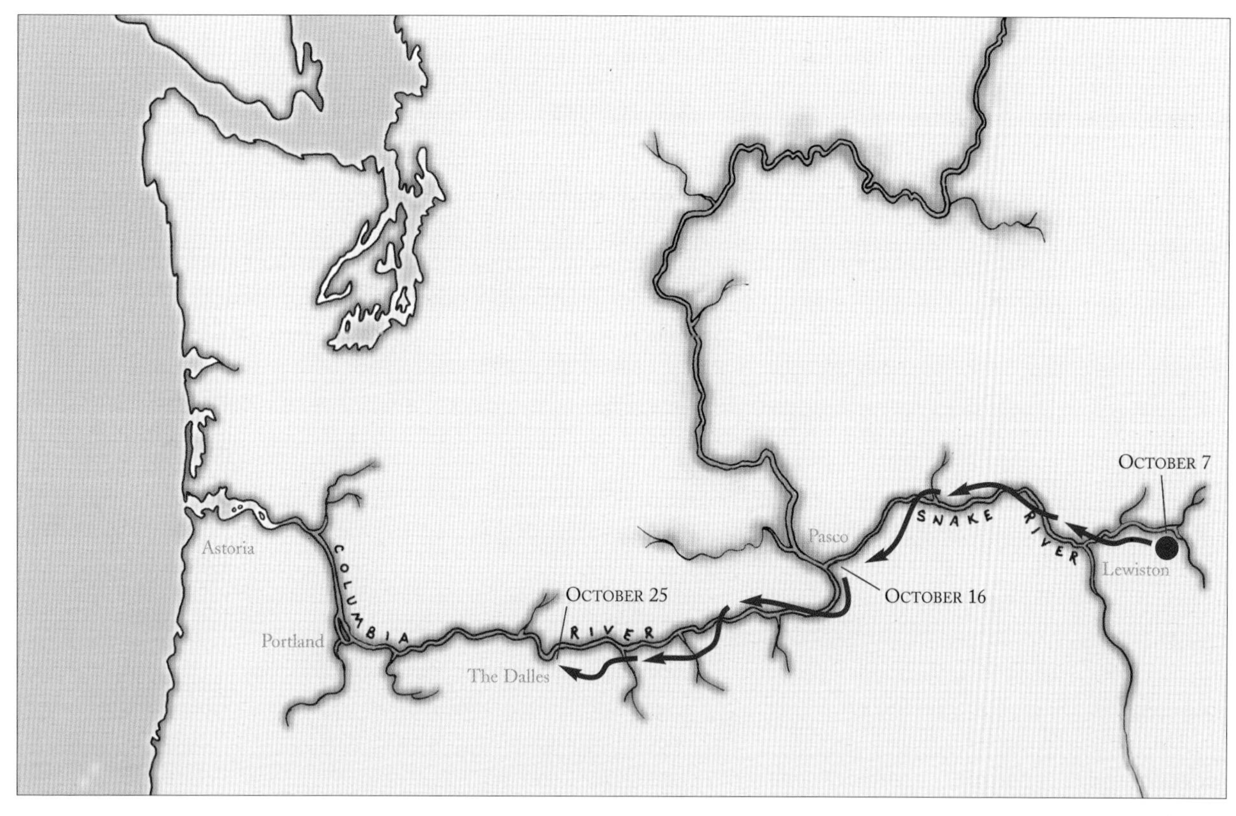

Lewis and Clark's party cross the Bitterroot Mountains on foot and arrive at the west-flowing Clearwater River, where they make five canoes. On October 7, they launch. After nine days of travel they reach the Columbia River; a week later they are encamped at the eastern side of the Cascade Range.

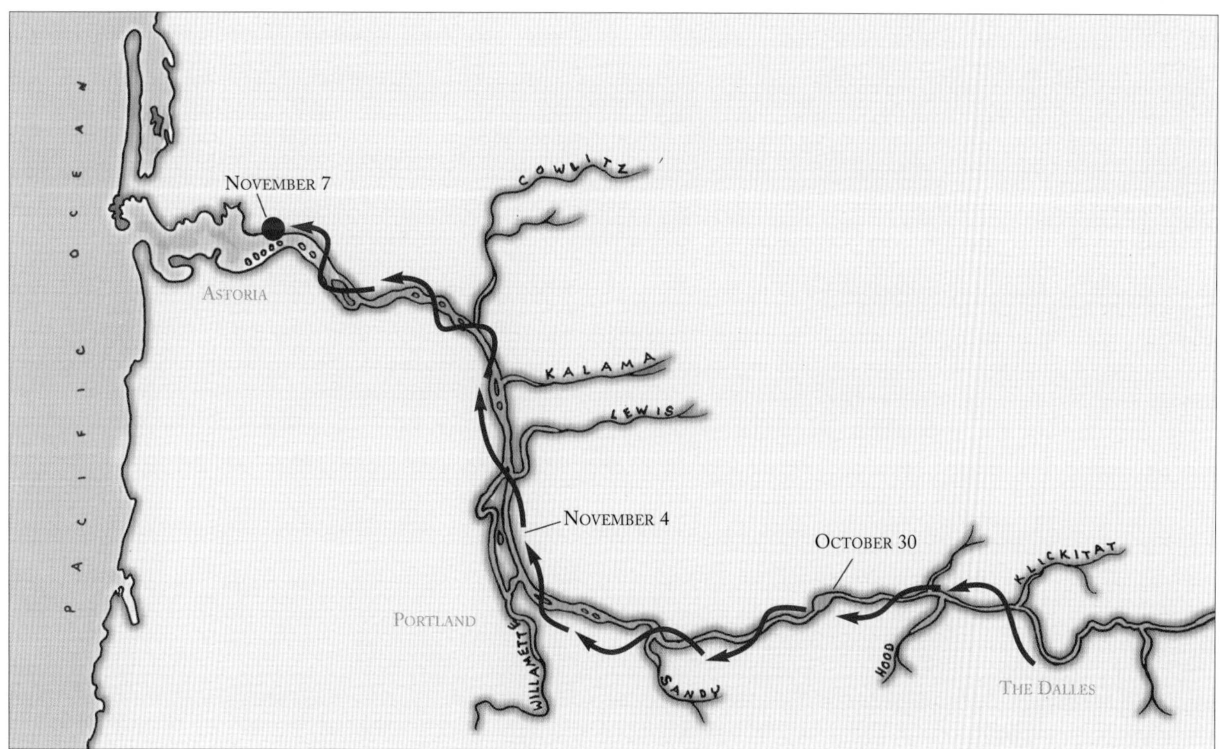

On the western side of the Cascades, the Columbia suddenly broadens into a gentle river nearly one mile wide. Travel is easy, and the party makes 30 miles a day. On November 7 exactly one month after beginning their water journey, they arrive in the lower Columbia estuary.

CHAPTER ONE

The Final Miles

Great joy in camp we are in View of the Ocian

Thursday, November 7th

Overcast skies and rain showers dominate the weather. A small low pressure storm moves towards the coast.

Daytime Low Tide:	*6:53 am*	*2.4*
Daytime High Tide:	*12:53 pm*	*9.4*
Sunrise:	*7:05 am*	
Sunset:	*4:53 pm*	

If Meriwether Lewis had turned and taken five steps away from William Clark, the two men would scarcely have been able to see each other. The fog of the lower Columbia River is like that. It drifts across the water in clouds so thick that visibility is reduced to nearly zero. Most travelers, when confronted with such fog, would pause and wait for it to lift, but not Lewis and Clark. On this morning of November 7, as soon as it was barely daylight, they had their canoes loaded and launched, and the entire party set out downriver, continuing their route to the great Pacific Ocean. Clark described the morning in his journal with these few brief words:

> *A cloudy foggey morning Some rain. we Set out early . . . the fog So thick we could not See across the river*
>
> Gary E. Moulton, *The Journals of the Lewis & Clark Expedition* 6:31 (Clark)

In spite of the poor visibility, the captains probably planned to keep their five canoes close together and follow the flow of the current. Unfortunately, after they had gone only a couple of miles, one crew separated from the others, made a wrong turn, and before anyone realized it, they were gone, lost in the morning mist. As odd as it might seem, Clark's journal indicates that no one appeared worried about the missing men; they merely paddled on downriver, apparently confident they would meet up again somewhere.

The fog lifts.

Around noon the fog began to lift, and this gave the men their first glimpse of the surrounding landscape. Everyone was hoping to see the ocean. They knew it had to be very close.[1] However, what they saw were the same steep hills and mountain ridges that had bordered the river during the past several days. Whether they looked north or south, east or west, all they saw were mountains. Clark accurately described this view:

Captain George Vancouver's 1792 map of the Columbia River showed that all ships anchored in the deep water along the northern shore, which perhaps explains why Lewis and Clark decided to continue along that side of the river. They hoped to meet a sailing ship, so logically they would have wanted to be on the side of the river where the ships anchored.[2]

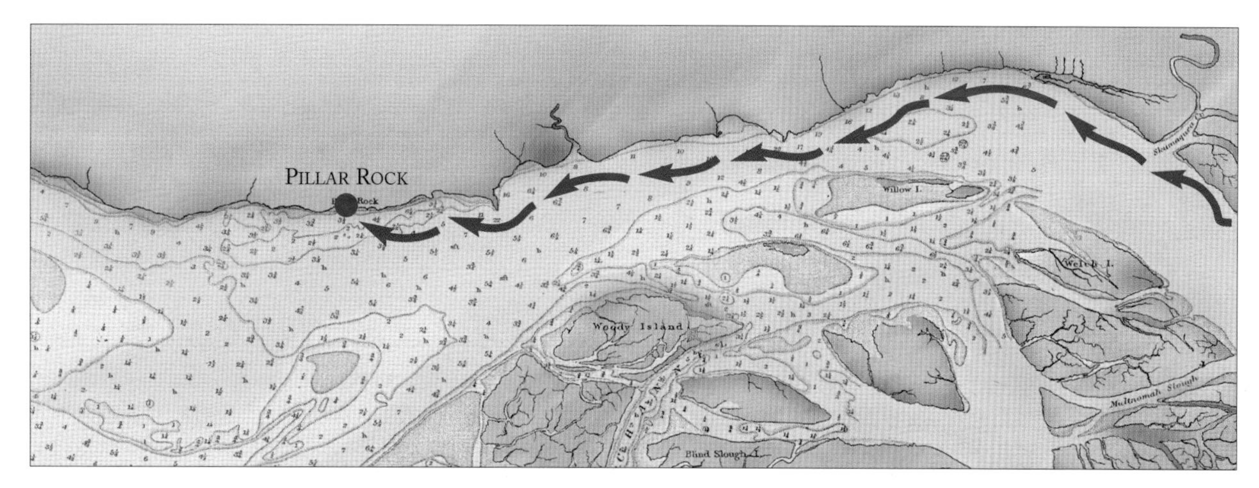

Lewis and Clark's party continue down river. They pass the final, sweeping curve and enter an enormous estuary crowded with dozens of islands.

As evening approaches they beach their canoes and set up camp for the night.

> *here the high mountanious Countrey approaches the river . . . a high mountn. to the SW. about 20 miles, the high mountans. Countrey Continue*
>
> Moulton, *Journals* 6:32-33 (Clark)

Directly in front of them, a steep ridge of hills cut right across their path like a huge natural wall and deflected the waters of the Columbia into a sharp curve.[3] Everyone could see that a dramatic change in the landscape was beginning to occur. The hills bordering the south shore diminished, and the river began to spread out wider than ever before. It grew from two miles in width to four, then doubled again. Clark noted this sudden change in the landscape:

> *the high mountaneous Countrey leave the river . . . the river widens into a kind of Bay & is Crouded with low Islands* Moulton, *Journals* 6:33 (Clark)

The party paddled along this great sweeping bend and proceeded downriver, keeping their canoes close along the right-hand shore. Everyone must have been wide-eyed, looking around at the scenery that was changing every minute. Towering spruce and cedar trees leaned precariously from the bank, their long, curved limbs swooping down nearly to the water. Thousands of ducks and geese filled the sky; their trumpeting and cries would have been heard for miles:[4]

> *here we See great numbers of water fowls about those marshey Islands* Moulton, *Journals* 6:32 (Clark)

The men reached forward with their paddles, sliced down into the water, and pulled back, stroke after stroke after stroke. Thirty days earlier they had begun their travels in these canoes, and by now the muscles in their arms and shoulders were finely conditioned for this precise motion. There was no confusion or bumping into one another; not a single paddle splashed carelessly into the water. Each crew worked in perfect synchronization, and in a typical day they could travel more than thirty miles.

By the time they had advanced downriver another several hours, the afternoon light began to fade, and it became time to find a campsite for the night. Generally this posed no problem, but here the shoreline was narrow and rocky. Good campsites did not exist. However, eventually they found a stretch of riverbank just large enough for them to spread out their blankets. The camp was far from perfect, but it would do for one night.

> *Shore boald and rockey and Encamped under a high hill . . . we with dificuelty found a place Clear*

Canoes

Lewis and Clark used five canoes to transport their party and supplies down the Columbia River. Four of these were rough-hewn dugouts the men had carved from tree trunks along the Clearwater River.[1] The fifth canoe was entirely different. It was a sleek, light Indian canoe that had been purchased from natives near Celilo Falls.[2] It also had been carved from a single log, but was sculpted specifically for travel through high, rough waves.

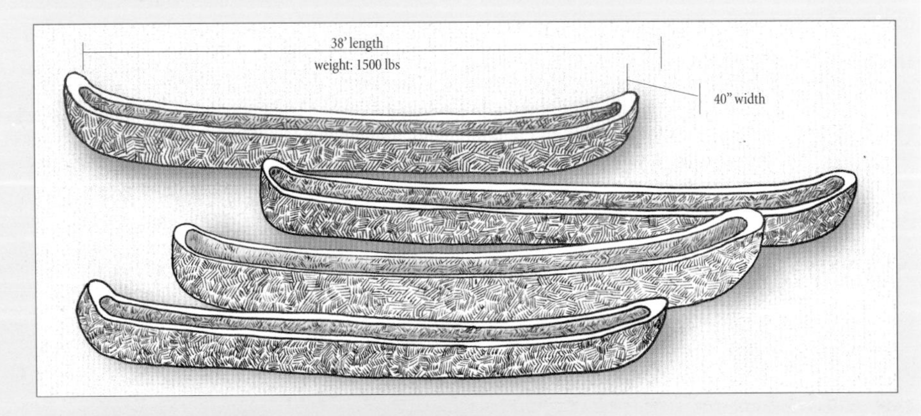

of the tide and Sufficiently large to lie on and the only place we could get was on round Stones on which we lay our mats rain Continud. moderately all day

Moulton, *Journals* 6:33 (Clark)

Fortunately, the lost crew suddenly appeared from behind an island. They had taken a different route but quickly paddled over to rejoin their companions. There was no doubt about it, Lewis and Clark always seemed to have good luck on their side.

our Small Canoe which got Seperated in the fog this morning joined us this evening

Moulton, *Journals* 6:33 (Clark)

The entire party was together once again, and they now found themselves in an environment the like of which they had never seen before. The river had grown into a massive body of water eight miles across. In fact, the distant shore was so far away that with the naked eye one could scarcely see any detail. Individual trees lost their shape and appeared like pastel green smudges. Farther back, the hills faded into frosty shades of blue.

the high hills on that Side, the river being too wide to See either the form Shape or Size

Moulton, *Journals* 6:33 (Clark)

The enormous river, the swirling clouds of geese, and the lush forest that bordered the water's edge created a breathtaking panorama, but the real excitement occurred when the men turned their attention downriver and gazed toward the west. Here they saw what they had long been hoping to see: the Columbia flowed

The southern shore of the wide Columbia is too distant to see from the north side of the river.

(below) Lewis and Clark look downriver and see the Pacific Ocean. They are now in view of their destination.

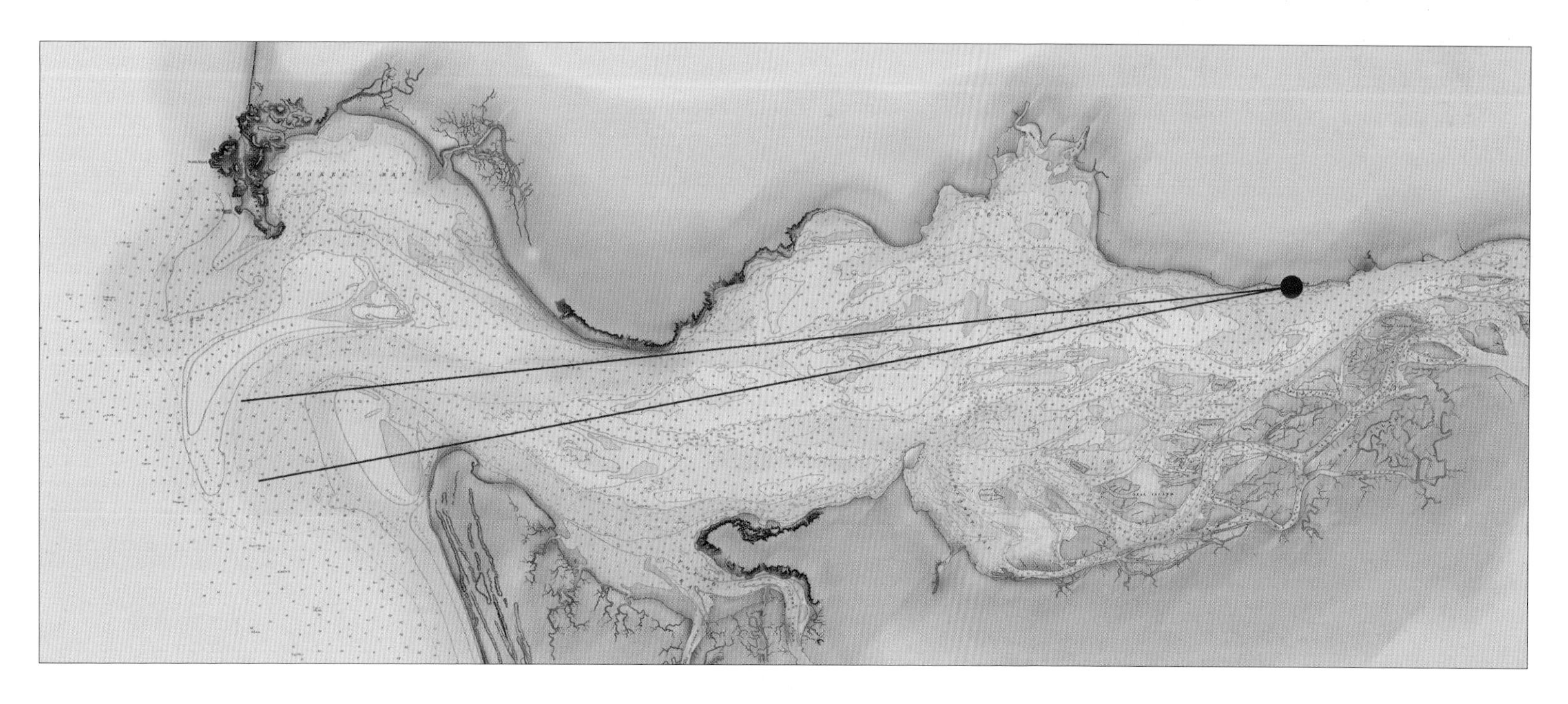

westward for another two dozen miles then disappeared over the horizon.

They must have been ecstatic. After eighteen months of difficult and dangerous travel, they had finally arrived at the far side of the continent. There was no question about it: from where they were camped they could see the mouth of the Columbia River and beyond *(see Appendix One: "To See or Not to See," page 188)*. Clark exclaimed:

> *Great joy in camp we are in View of the Ocian, this great Pacific Octean which we been So long anxious to See* — Moulton, *Journals* 6:33 (Clark)

Mouth of the Columbia: Graveyard of the Pacific

The mouth of the Columbia River is referred to as the *Graveyard of the Pacific* because of the thousands of wrecks and the horrific losses of life that have occurred there.[1]

What makes these particular waters so dangerous is a combination of many factors. The river is enormous, the current is strong, fog hovers low, and the sand bars shift their location invisibly beneath the waves. Adding to these hazards are the frequent winter gales that strike this region between October and April. The hurricane-force winds stir up the ocean into swells measuring twenty to forty feet in height which surge for miles up into the river. A pilot error or mechanical malfunction can cause a ship to run aground where it is then torn apart by the current and waves.

In 1811 Ross Cox approached the Columbia River aboard a ship and described what he saw:

> *The mouth of Columbia River is remarkable for its sand-bars and high surf at all seasons, but more particularly in the spring and fall, during the equinoctial gales: these sand-bars frequently shift, the channel of course shifting along with them, which renders the passage at all times extremely dangerous. The bar, or rather the chain of sand-banks, over which the huge waves and foaming breakers roll so awfully, is a league broad, and extends in a white foaming sheet for many miles, both south and north of the mouth of the river, forming as it were an impracticable barrier to the entrance, and threatening with instant destruction everything that comes near it.*[2]

This was a moment the two officers would never forget. Captain Meriwether Lewis, only thirty-one years of age, and Captain William Clark, four years his senior, had just succeeded in doing what no one before them had ever done: they had led the first group of Americans across the continent.[5] The end of their westward journey seemed now within easy reach.

As darkness arrived, the men wrapped themselves in their blankets and bedded down for the night. Even the obnoxious rain wasn't enough to dampen the sense of joy and relief they felt. Tomorrow they would paddle this final stretch of river, set up camp near the ocean, and be finished with their journey. By all appearances, this day of travel was going to be easy.

However, if this is what they were assuming, they could not have been more mistaken. From their campsite, it appeared as though the river spilled into the ocean without so much as a ripple, like warm molasses pouring from a barrel. What the men had no way of knowing was that the mouth of this great river, where its waters collide head-on with the Pacific Ocean, was a nightmarish place where huge breakers churned from shore to shore. Even experienced sea captains in sturdy ships dreaded this place. In fact, the mouth of the Columbia River eventually earned the reputation of being one of the most dangerous and unpredictable bodies of water in the world.

For now, Lewis and Clark's men were completely unaware of any threat. They heard the constant drumming of the distant surf, and this reassured them that by tomorrow they would get their first close-up look at this great western ocean.

> *the roreing or noise made by the waves brakeing on the rockey Shores . . . may be heard distictly*
>
> Moulton, *Journals* 6:33 (Clark)

Unfortunately, none of Lewis and Clark's men had any idea that the worst part of their entire expedition was still to come. That distant roaring noise they heard was the sound of danger, and they would be heading directly towards it.

Major Ship Wrecks 1829 – 1936

1. Marie	1852	20. U S Grant	1871	39. Lewis	1865	58. Cairnsmore	1883
2. Klose	1908	21. Morning Star	1849	40. Maine	1848	59. Peter Iredale	1906
3. Jennie Ford	1864	22. Nimbus	1877	41. Kake	1913	60. Great Republic	1879
4. Gov. Moody	1890	23. Corsica	1882	42. Mindora	1853	61. Windward	1871
5. Vandalia	1883	24. Edith Lorne	1881	43. Delharree	1880	62. Harvest Home	1882
6. Rosecrans	1913	25. Cape Wrath	1901	44. J. Merithew	1853	63. Washington	1911
7. Laurel	1929	26. USS Peacock	1841	45. W.B. Scranton	1866	64. Walpole	1849
8. Iowa	1936	27. Sidi	1874	46. J.C. Cousins	1883	65. Gamecock	1898
9. Admiral Benson	1930	28. Aurora	1849	47. Sloop	1856	66. Sulphur	1839
10. Pescawha	1933	29. Potomac	1852	48. Devonshire	1884	67. W.H. Besse	1886
11. Cavour	1903	30. Admiral	1912	49. Fern Glen	1881	68. Jenny Jones	1864
12. Champion	1870	31. Ariel	1886	50. Primrose	1882	69. Allegiance	1879
13. Ellen	1870	32. Vancouver	1848	51. Oshkosh	1911	70. Detroit	1855
14. Light Ship #50	1899	33. Sea Thrush	1932	52. Oriole	1853	71. J. Merithew	1853
15. Rival	1881	34. Josephine	1849	53. Woodpecker	1861	72. Enterprise	1858
16. North Bend	1928	35. Rochelle	1914	54. General Warren	1852	73. Bordeaux	1852
17. William Reese	1886	36. William & Ann	1829	55. City of Dublin	1878	74. Dreadnaught	1876
18. Alsternixe	1903	37. Industry	1865	56. Dolphin	1852	75. Desdemona	1857
19. Isabella	1830	38. Shark	1846	57. Architect	1875		

Courtesy of the Columbia River Maritime Museum

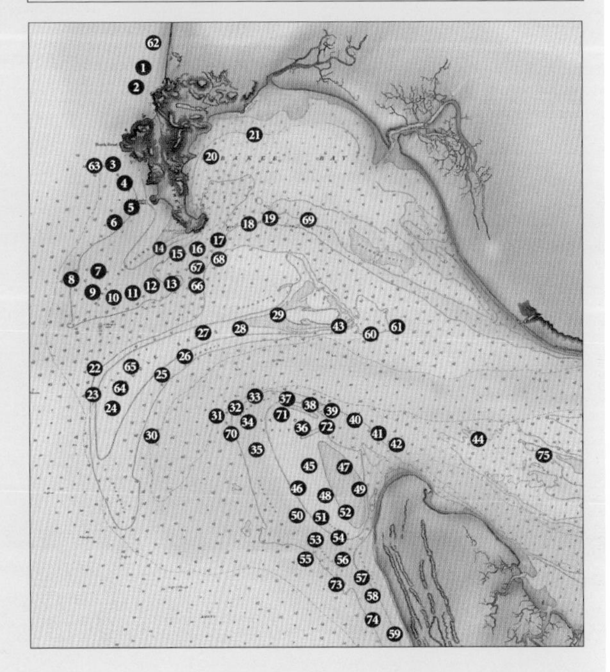

Friday, November 8th

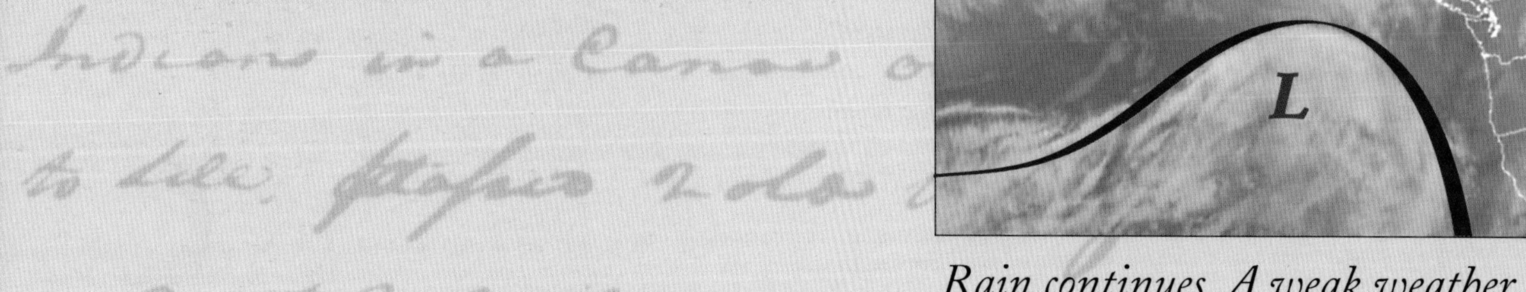

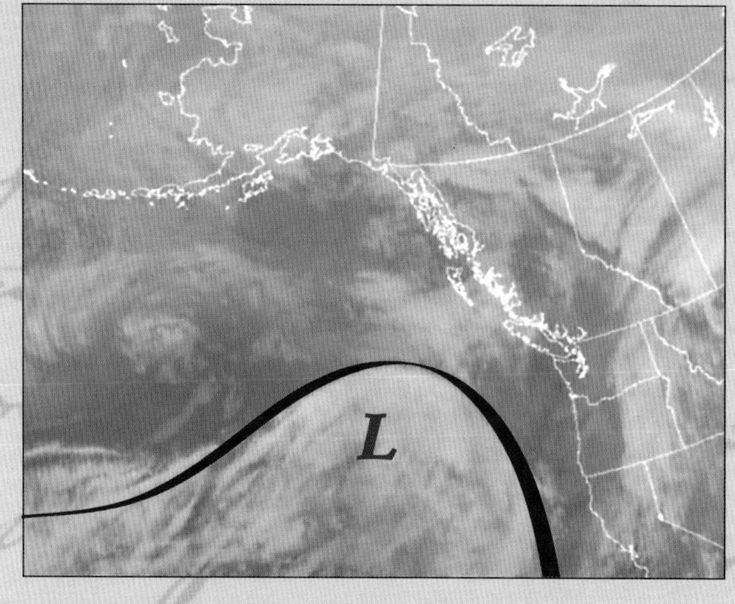

Rain continues. A weak weather system approaches the coast.

Daytime Low Tide:	*7:35 am*	*2.7*
Daytime High Tide:	*1:33 pm*	*9.5*
Sunrise:	*7:06 am*	
Sunset:	*4:53 pm*	

THE THICK FOG DID NOT RETURN DURING THE night. Dark gray clouds drifted across the sky, but the visibility was good. Captain Clark, who was an experienced cartographer, could have looked down the river and estimated that the Pacific Ocean was scarcely twenty-five miles away, a distance they could easily travel in six hours. If they set out immediately, they could be standing on the beach by early afternoon.

Here this history takes a peculiar turn. Lewis and Clark did not launch their canoes. Instead, the men lingered around camp for several hours and changed their clothes. There's no explanation in the journals of why they did this; but since it was a tradition in those days to arrive at a destination dressed in one's finest clothes, it appears that the men were grooming themselves for their arrival at the ocean.

> *A Cloudy morning Some rain, we did not Set out untill 9 oClock, haveing Changed our Clothing*
>
> Moulton, *Journals* 6:35 (Clark)

By the time they finished these preparations and set out, the morning was half gone. The men paddled their canoes along the shore, which ran straight for several miles then abruptly ended at the upper tip of a bay. Here Lewis and Clark were faced with two choices: either cut straight across and continue on downriver, or turn and lead their canoes along the perimeter. Cutting straight across would be the quickest and most direct route, but this choice would expose them to miles of open water. It was much safer to keep their heavily loaded canoes close to shore.[6]

> *at 3 miles entered a nitch of about 6 miles wide and 5 miles deep* Moulton, *Journals 6:36* (Clark)

Change of Clothes

There was a tradition among long-distance voyagers to commemorate the end of a journey by dressing up in their best clothes for the final day of travel. For example, in 1824 Governor George Simpson of Hudson's Bay Company described such preparations after a journey lasting eighty-four days.

> *After Supper all hands were busily occupied in shaving scrubbing and changing as by continuing our march during the night we expected to reach Fort George the following morning*[1]

Frederick Merk, professor of history at Harvard University, explained this tradition.

> *It was a custom among voyageurs to end any extended journey with a flourish. . . . Preparations consisted of scrubbing and packing away the soiled clothing worn on the journey and donning in its place show garments, including a flashy sash tied about the waist, ribbons braided into the hair, and moccasins embroidered with gay beadwork. Generally a stop was made for the purpose of "changing" shortly before reaching the point of destination*[2]

Narcissa Whitman, one of the first two pioneer women to cross the Rocky Mountains, described their evening before arriving at Fort Walla Walla in 1832. The party had been traveling for six and a half months and was weary of the trail, but nevertheless they halted before approaching their destination.

> *Our employment this afternoon is various. Some are washing their shirts & some are cutting their hair others shaving, preparing to seeing Walla Walla . . . It is the custom of the country to send heralds ahead to announce the arrival of a party and prepare for their reception.*[3]

Exactly how much of Lewis and Clark's clothing had worn out at this point in their expedition is open to debate. We know that some men were wearing buckskin clothing, but it is entirely possible that they still possessed portions of their uniforms.[4] Clark, unfortunately, does not provide any details.

The captains may have brought along regimental uniforms to be used for special occasions. Likewise, it is possible the three sergeants and some of the privates also still retained pants, shirts, coats, sashes, or hats from the military uniforms issued to them.

Captain Meriwether Lewis *Captain William Clark*

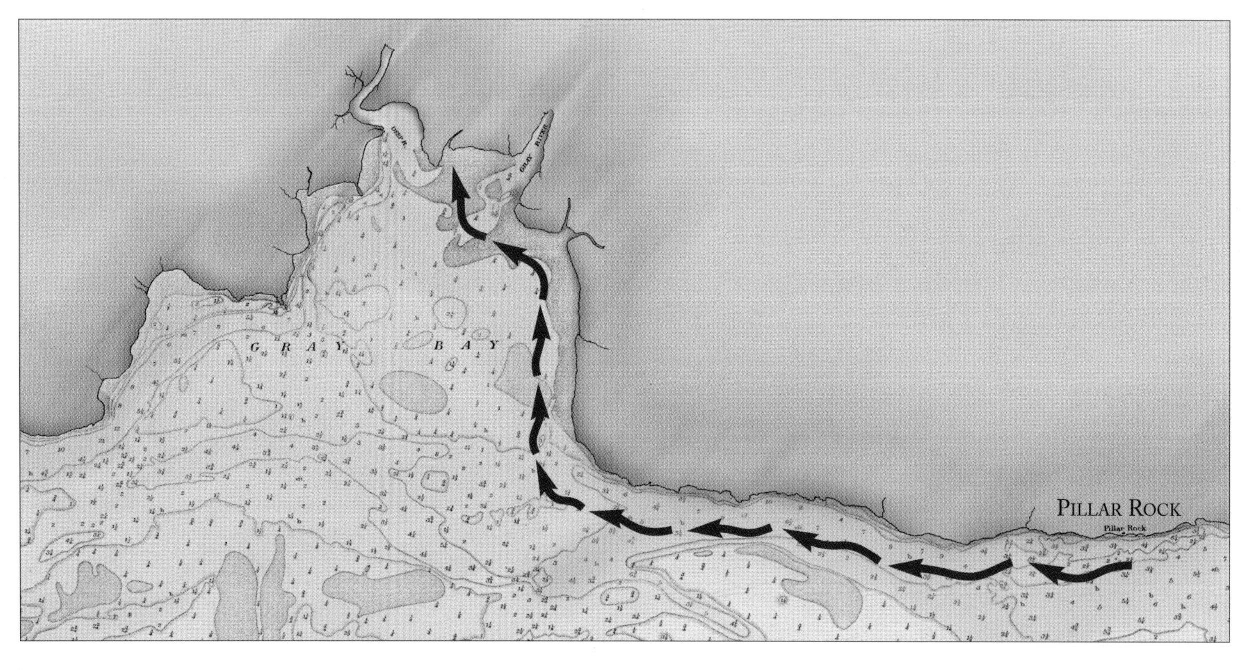

The party proceeds downriver along the shore until they arrive at a large bay. Here, rather than risking crossing the open water, they turn and follow along its shoreline. Finding a convenient place to stop for a midday meal, the party lands and waits until the tide changes.

Throughout the entire morning, the Columbia's current had been flowing upriver. Paddling against this current caused extra work and was entirely unnecessary. The captains knew that if they paused for a couple of hours, the tide would turn and they could use the ebbing current to their advantage, zipping down to the ocean at twice their normal speed.[7] An unoccupied Indian village came into view, so they decided to pull ashore and wait. Besides, it was time to eat breakfast.

> *we came too at the remains of an old village at the bottom of this nitch and dined . . . Sent out 2 men and they killed a Goose and two Canves back Ducks*
>
> Moulton, *Journals* 6:36 (Clark)

As the men ate their breakfast, the tide filled the vast river. The current slowed down, reached high water, then turned and began to flow back to the ocean. It was time to leave. All they had to do was paddle out into the river's main channel, and be swept down to the ocean by the great ebbing flow.

> *we took the advantage of a returning tide and proceeded on to the Second point*
>
> Moulton, *Journals* 6:36 (Clark)

They had launched their boats hundreds of times since they began this journey back in May 1804, but this was the most exciting and memorable launching because it would be the last. The next time they stepped out of these canoes they would be standing directly in front of the Pacific Ocean – or so they thought.

The canoes slipped along the shore of the bay, passed one point of land and proceeded on to the next. The waters were calm and trouble-free.

However, when they rounded the point where the waters of the bay met the open river, Lewis and Clark received an unwelcome surprise. Here they ran directly into whitecaps created by a combination of wind and ocean surge. Their canoes plunged deeply into each wave, then dipped and rolled and reared up like bucking horses. Water splashed into the air and tossed the crew around as they fought to keep their canoes

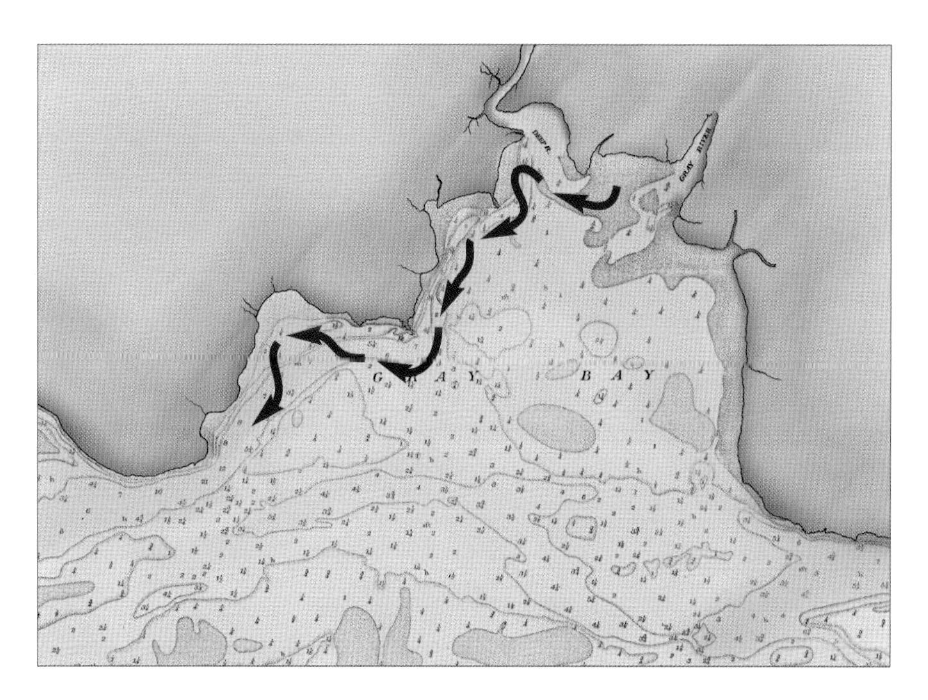

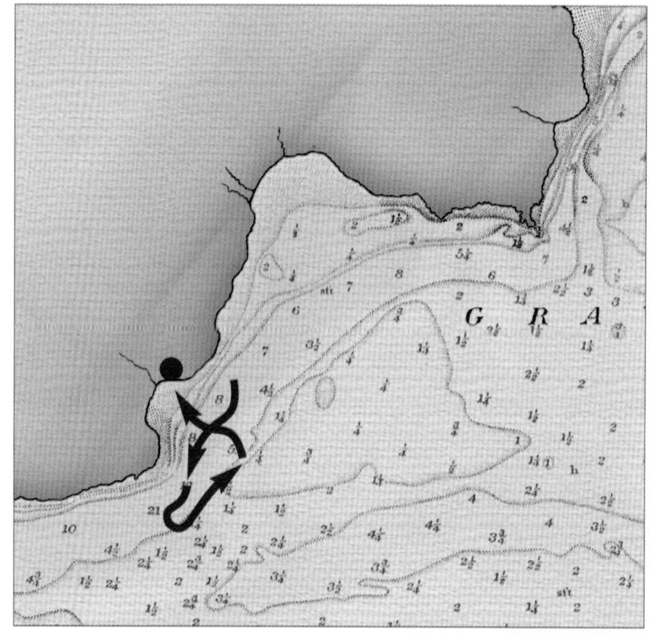

(above left) After high water slack the ebb begins. Lewis and Clark's men launch their canoes and continue on downriver.

(above right) The party unexpectedly encounters large waves. Wisely, they turn around and beach their canoes.

nosed into the waves. The violent motion nauseated several members of the party.

> *The Swells were So high and the Canoes roled in Such a manner as to cause Several to be verry Sick. Reuben fields, Wiser McNeal & the Squar*
>
> Moulton, *Journals* 6:35 (Clark)

These rough waters brought Lewis and Clark's progress to a complete stop. It was too dangerous to continue, so they turned and landed along a nearby tip of shore.

> *here we found the Swells or waves So high that we thought it imprudent to proceed; we landed unloaded and drew up our Canoes*
>
> Moulton, *Journals* 6:36 (Clark)

They intended to remain only briefly, perhaps to rearrange their loads so the canoes would be balanced in the rough waters. However, before they could set out again, the waves grew in size and rolled from shore to shore, making it virtually impossible to proceed.

> *the waves are increasing to Such a hight that we cannot move from this place, in this Situation we are compelled to form our Camp*
>
> Moulton, *Journals* 6:36 (Clark)

Now they were stuck. To make matters worse, this stretch of shoreline was a narrow sliver of rock covered with huge driftwood logs; behind this was a near-vertical hillside. In other words, there was absolutely no place to lie down. This was the worst possible campsite they could have chosen.

> *our present Situation a verry disagreeable one in as much; as we have not leavel land Sufficient for an encampment and for our baggage to lie Cleare of the tide, the High hills jutting in So Close and Steep that we cannot retreat back* Moulton, *Journals* 6:36 (Clark)

Another annoyance was the persistent rain. Most of the men had no cover from the downpour. Some of them set to work building large fires to dry themselves, while others may have found shelter by crawling in beneath the massive driftwood logs. It wasn't comfortable, but they could endure it until morning.

White men

Lewis and Clark use the term "white men" to refer to non-natives. The Indians called the foreigners by different names. Most Americans were called *Boston Men* whereas the sailors from England were known as *King George Men.* Lewis tells us the Clatsop called them *pah-shish-e-ooks* which translates as *cloth men.*

That night, Lewis and Clark's conversation turned to the mysterious fur traders who reportedly lived near the ocean. Some Indians had indicated that these "white men" were permanent residents whereas others described them coming and going aboard ships. The Wahkiakums, in fact, had said that the fur traders were gone. Clark's uncertainty about whether or not there were any fur traders around the river's mouth was well expressed in his journal.[8]

> *We are not certain as yet if the whites people who trade with those people . . . are Stationary at the mouth, or visit this quarter at Stated times for the purpose of trafick* Moulton, *Journals* 6:36 (Clark)

They desperately wanted to meet anyone who could replenish their supplies. Though it was not yet a matter of life or death, they were in short supply of many essentials, and this would be their only opportunity to obtain more.

Twenty-four hours earlier, Lewis and Clark had thought they would be at the ocean by now; instead, here they were stuck along the shore, still a dozen miles shy of their destination. It was a shock and a disappointment.

That evening, they reloaded their baggage into the canoes and got everything ready for a speedy departure in the morning. They would set out from this miserable place as soon as the waters grew calm.

The Currents of the Lower Columbia River

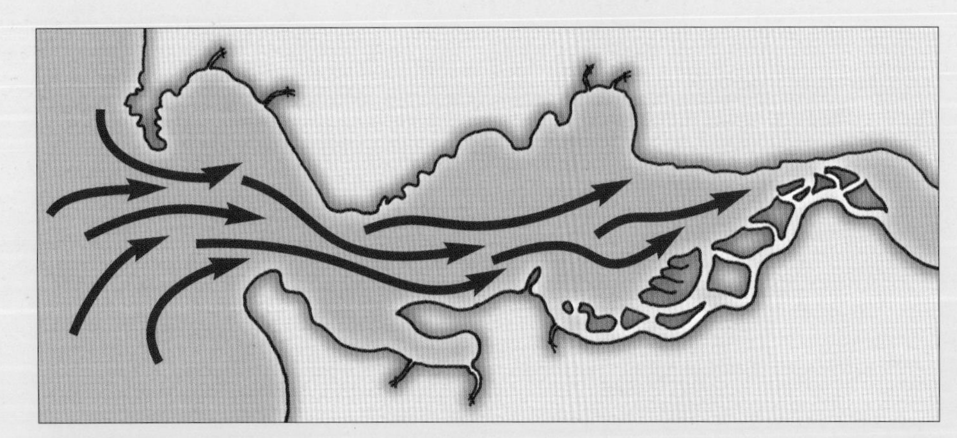

flood tide

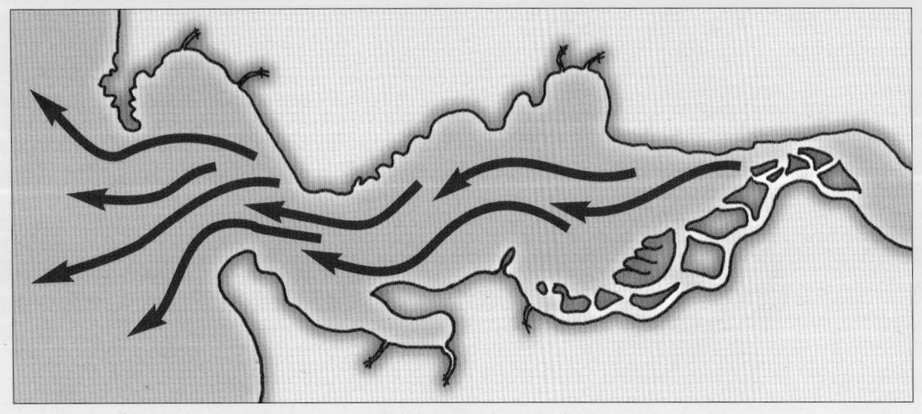

ebb tide

From the beginning of their journey Lewis and Clark's party had traveled on rivers that constantly flowed in the same direction, day after day. However, in the lower Columbia they were faced with tidewater, which presented them with an entirely different situation.

During a **flood tide** the ocean pushes into the Columbia, forcing the current to flow upriver. The water moves from west to east.

After **highwater** the **ebb tide** begins. The ocean recedes, the river current reverses and it flows out into the ocean. At this time the current moves from east to west.

Every six or seven hours the current reverses its direction. This motion of the water affects anyone in a boat; traveling "with the current" can double one's speed. Because of this, the fastest way to get somewhere by boat often requires waiting for the tide to flow in the same direction you intend to travel.

Rumors of White Men

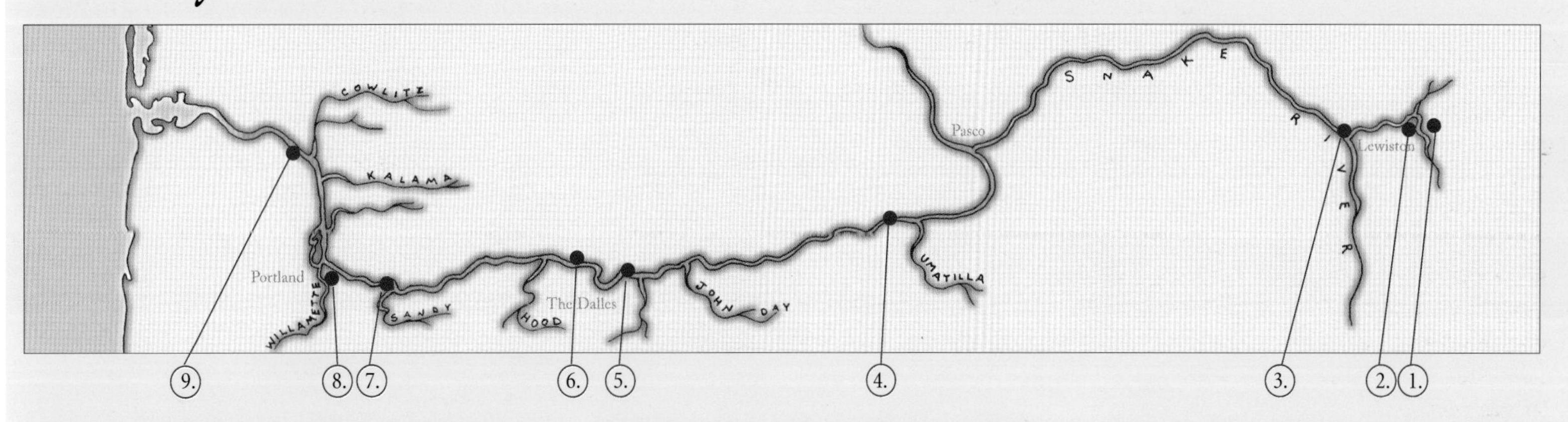

Indians who lived in the Bitterroot Mountains began to tell Lewis and Clark stories of white men who resided along the Columbia and near the ocean.[1] The information was never specific, but it was repeated often.

1. *one of them Drew me a Chart of the river & nations below informed of one falls below which the white men lived from whome they got white beeds cloth &c.*[2] (Clark)

2. *one old man informed us that he had been to the White peoples fort at the falls & got white beeds &c.*[3] (Clark)

3. *here we met with an Indian from the falls at which place he Sais he Saw white people.*[4] (Clark)

4. *We here saw some articles which shewed that white people had been here or not far distant during the summer. They have a hempen seine and some ash paddles which they did not make themselves.*[5] (Gass)

5. *We conclude that their must have been some white people among these Indians, as they had among them, a new Copper Tea kettle, beads, small pieces of Copper & a number of other articles We saw also a Child among them, which was a mix'd breed, between a White Man & Indian Women. The fairness of its Skin, & rosey colour, convinced us that it must have been the case, and we have no doubt, but that white Men trade among them.*[6] (Whitehouse)

6. *we call this the friendly village. I observed in the lodge of the Chief Sundery articles which must have been precured from the white people, Such a Scarlet & blue Cloth Sword Jacket & hat.*[7] (Clark)

7. *we met 2 Canoes, of Indians 15 in number who informed us they had Seen 3 Vestles 2 days below us*[8] (Clark)

8. *The Indians at the last village have more Cloth and uriopian trinkets than above I Saw Some Guns, a Sword, maney Powder flasks, Salers Jackets, overalls, hats, & Shirts*[9] (Clark)

9. *we over took two Canoes of Indians going down to trade one of the Indians Spoke a fiew words of english and Said that the principal man who traded with them was Mr. Haley... he Showed us a Bow of Iron and Several other things which he Said Mr. Haley gave him.*[10] (Clark)

The first foreigners to have contact with West Coast Indians were fur traders and early sailors. Pictures of these men do not exist; however, by looking at photographs of sailing crews from the 1890s, we can begin to imagine the types of men that the Indians first met.

Saturday, November 9th

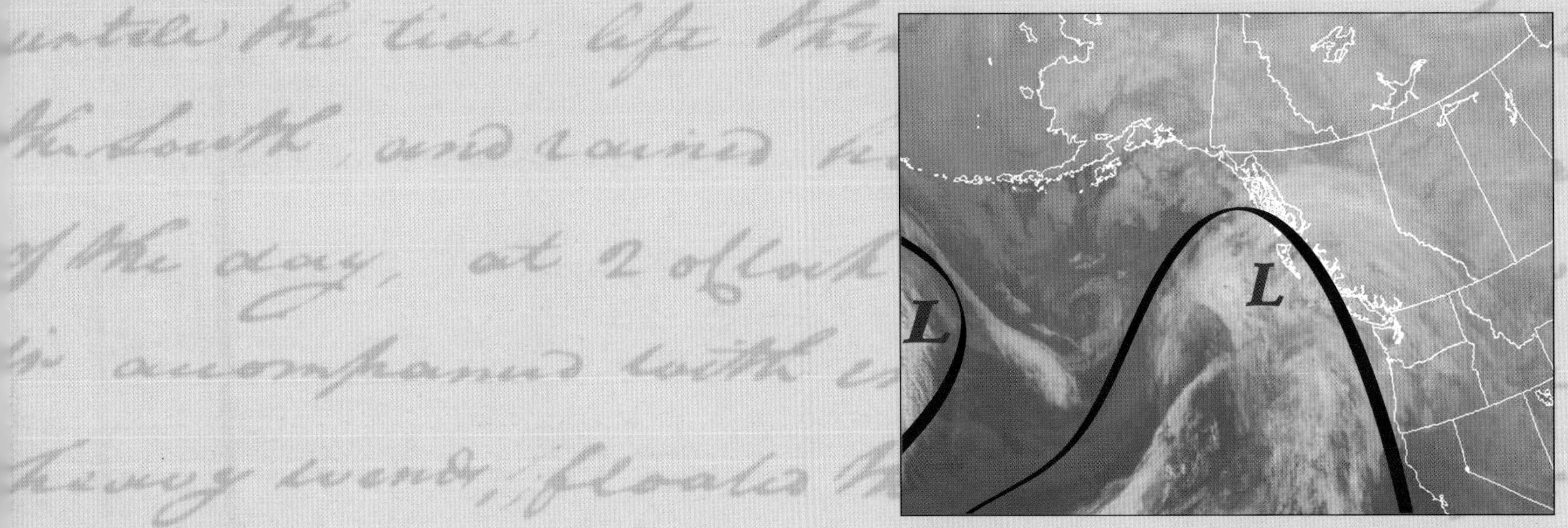

Hard rain and gusty wind dominate coastal weather. The storm moves north into Canada; another low pressure system builds far out in the Pacific Ocean.

Daytime Low Tide:	*8:21 am*	*2.9*
Daytime High Tide:	*2:17 pm*	*9.4*
Sunrise:	*7:07 am*	
Sunset:	*4:52 pm*	

Lewis and Clark's decision to leave their canoes loaded with baggage turned out to be a big mistake. That night the rising tide threatened to swamp their canoes and float away their precious supplies.[9]

Fortunately, one of the men noticed what was happening and sounded the alarm; quickly, everyone flew to work unloading the canoes. Clark's journal does not describe how the men were able to run around in the darkness over the black, slippery rocks in the pouring rain without falling and breaking bones, but they did it, and their supplies were successfully rescued.[10]

> *The tide of last night obliged us to unload all the Canoes one of which Sunk before She was unloaded by the high waves or Swells which accompanied the returning tide, The others we unloaded*
>
> Moulton, *Journals* 6:37 (Clark)

The rest of the night passed without incident. However, within the first few moments of daylight Lewis and Clark saw the very opposite of what they had been hoping to see. The waters of the Columbia had actually grown more threatening. Looking out from their camp, they could see that it would be suicidal to launch their heavy canoes in such rolling, breaking waves.

> *wind Hard from the South and rained hard all the fore part of the day*
>
> Moulton, *Journals* 6:38 (Clark)

The party was stranded, and there was nothing they could do but wait. Some men tried to venture out with rifles to shoot something to eat, but most of the crew huddled close to the huge bonfires, first warming one side of their bodies while the rain soaked and chilled the other, then turning to warm and dry the other side. It was an endless process of getting wet, then dry, and then wet once again. Obviously, this was going to be a long and unpleasant day.

Meanwhile, completely unknown to the men, trouble was on its way from downriver. The flood tide had begun once again. Millions and millions of gallons of water from the ocean rushed into the

When Indians Camp

Several months later Lewis noted that the Indians never left their canoes loaded at night.

> *when the natives land they invariably take their canoes on shore, unless they are heavily laden, and then even, if they remain all night, they discharge their loads and take the canoes on shore.*[1]

Lewis and Clark's party are only 12 miles away from the ocean and yet unable to move.

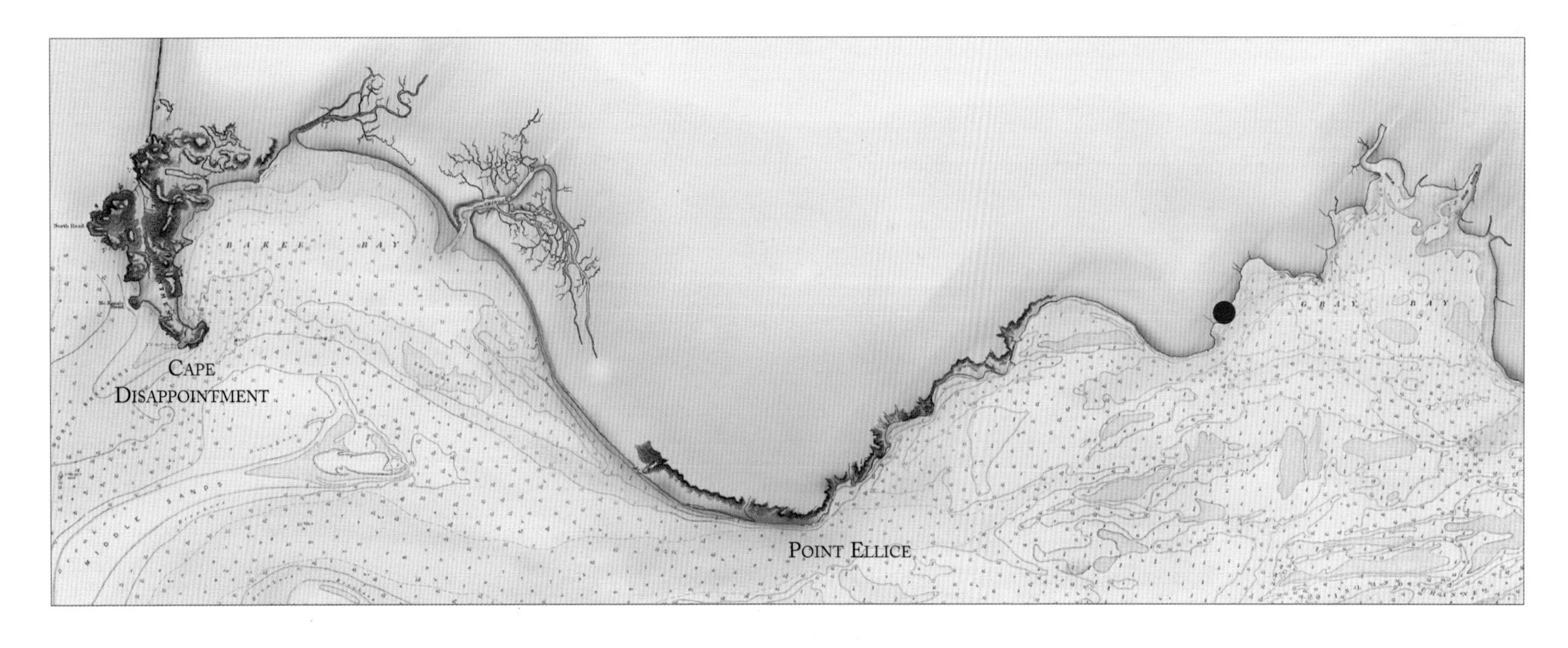

"with every exertion and the Strictest attention by every individual of the party was Scercely Sufficient to Save our Canoes from being crushed by those monsterous trees"

William Clark

Tidewater of the Lower Columbia River

During high tide some ships became stranded in shallow water. When this happened the local townspeople would amuse themselves during low water by taking photos and visiting with the crew. Often on the next high tide the ship would refloat and be pulled out to sea.

The waters of the Lower Columbia River move up and down with every tide. The average drop or rise is eight and a half feet; however, during extreme tides the water level can move as much as eleven vertical feet.

This tidewater created a terrible nuisance for Lewis and Clark's party. If they camped too close to the river, the rising tide would come right into their camp, extinguish the fires, and soak their blankets; a canoe left unattended would float away. On the other hand, if they pulled their heavy canoes ashore during a high tide, soon they were hundreds of yards away from the water.

mouth of the Columbia, causing the level of the river to rise at the rate of one inch every four minutes.[11] While they were traveling in canoes, the changes in river level were barely perceptible and caused them no problems, but now that they were stuck along the shore it had a devastating effect. Slowly at first, the shoreline began to disappear; four hours later, the water had risen more than six feet. The waves smashed into their bonfires with one loud hiss and turned the red-hot coals into black lumps.

The waters herded the men together and pressed them back against the driftwood logs that lay along the shore. Moments later, the rising tide was around their ankles and grew deeper by the minute. With no other place to go, they climbed up on top of the massive logs. Here they would be safe, or so they thought.

Soon the water had risen so high that it completely surrounded the stranded driftwood. And then, to Lewis and Clark's horror, the logs began to move. As unbelievable as it must have seemed, these enormous tree trunks, some of them with diameters wider than a barn door is tall, began to roll and bounce with the surge of every wave.

> *at 2 oClock P M the flood tide came in accompanied with emence waves and heavy winds, floated the trees and Drift which was on the point* Moulton, *Journals* 6:38 (Clark)

This rapidly became an extremely dangerous situation. The rolling logs would eventually throw all of their supplies into the water, so the men wrestled their baggage farther up the steep slope and stashed it behind trees and logs. From this perch they could look down at what had once been their campsite, now completely submerged.

While all of this was occurring, the rain continued to pour down without pause, and the wind blew gusts of thirty to fifty miles per hour. The combination of being soaking wet and whipped by strong wind must have felt absolutely dreadful. It's hard to imagine the men enduring such hardship, but it is even more difficult to imagine young Sacagawea and her little nine-month-old baby suffering through this ordeal.

The massive driftwood logs that were strewn along the shore were now all afloat, and the tremendous power of the Columbia River pushed these logs into the shore with shattering force. The logs smashed into each other like a giant rock crusher, then spun and rolled on the surge of every wave. Suddenly, a new crisis emerged. Their canoes became entangled in this heaving mass of driftwood, and everyone could see that it was just a matter of time before the canoes would get pinched between the logs and split beyond repair.

The thought of making more canoes would have sent a shiver up everyone's spine. Where could they find a place to do this along the vertical hillside; how would they deal with the tides, and how many exhausting weeks of work would it require? It was obvious that they had no choice other than to save these canoes, and there wasn't a minute to spare. The men nimbly jumped from log to log, and pushed and pulled against the floating tangle of trees to free their canoes. The logs slammed and rolled against one another with such force that if one of the men were to slip and fall in between, he would be crushed to death in an instant. Clark described the waves.

> *tosed them about in Such a manner as to endanger the Canoes verry much, with every exertion and the Strictest attention by every individual of the party was Scercely Sufficient to Save our Canoes from being crushed by those monsterous trees maney of them nearly 200 feet long and from 4 to 7 feet through* Moulton, *Journals* 6:38 (Clark)

The rising tide overwhelms their camp.

Wind Chill

Lewis and Clark's men were soaking wet and dressed in worn-out clothing. The temperature they experienced would have felt much lower than that represented on this chart.

Wind Speed (mph)	Air Temperature			
	50°	45°	40°	35°
20	32°	25°	18°	11°
25	30°	23°	16°	8°
30	28°	21°	13°	5°
35	27°	19°	11°	3°
40	26°	18°	10°	2°
45	25°	17°	9°	1°

The winds that accompany coastal storms range in speed from twenty to as much as one hundred miles per hour. This wind, when combined with the average temperature, (somewhere between 35° to 48°) can feel numbing cold.[1]

The rain poured down by the bucketful, and the wind whipped their already tattered clothes into rags. There was nothing to eat except dried fish and no chance to make a fire for warmth. Everyone was cold, hungry, and soaked to the skin.

> *our camp entirely under water dureing the hight of the tide, every man as wet as water could make them all the last night and to day all day as the rain Continued all day*
> Moulton, *Journals* 6:38 (Clark)

Finally, the ebb began. Inch by inch the water receded. One rock emerged, then another and another. The huge bouncing, rolling driftwood logs came to rest like a herd of dinosaurs lying down to sleep. The small shoreline emerged once again, and finally the men could climb down from their hillside perch. It seemed as if the worst was over.

Then suddenly, the full force of the storm hit. The wind whipped up and drove the rain sideways. The men had to lean into the gusts to stand upright. The bonfires had to be delayed again for several hours, which added to their misery.

> *at 4 oClock P M the wind Shifted about to the S.W. and blew with great violence imediateley from the Ocian for about two hours*
> Moulton, *Journals* 6:38 (Clark)

As nighttime approached, William Clark described the mood of his men as being upbeat. It is hard to imagine that after such an ordeal the men would be in good humor, but they apparently were looking forward to their imminent arrival at the ocean.

> *notwithstanding the disagreeable Situation of our party all wet and Cold . . . they are chearfull and anxious to See further into the Ocian*
> Moulton, *Journals* 6:38 (Clark)

It had been a grueling day. The storm had given them a big scare and caused them hours of dangerous toil. They didn't need any more problems, yet one more ordeal awaited them. Some of the men who hungered for salt had ingested the briny ocean water during the high tide, and this salt charged through their digestive systems like a powerful laxative.[12] It is difficult to say which group suffered more, those with diarrhea or those who had to share the same narrow beach. Clark sums up the evening with these words:

> *at this dismal point we must Spend another night as the wind & waves are too high to proceed*
> Moulton, *Journals* 6:38 (Clark)

Fortunately, the storm quickly passed and the wind grew calm, which gave Lewis and Clark every reason to believe that tomorrow they would finally arrive at the ocean. The men curled up in their blankets, wet, exhausted, and hungry, but ready to set out as soon as it was safe to launch their canoes.

CHAPTER TWO

STRUGGLE AROUND POINT DISTRESS

It would be distressing to a feeling person

Sunday, November 10th

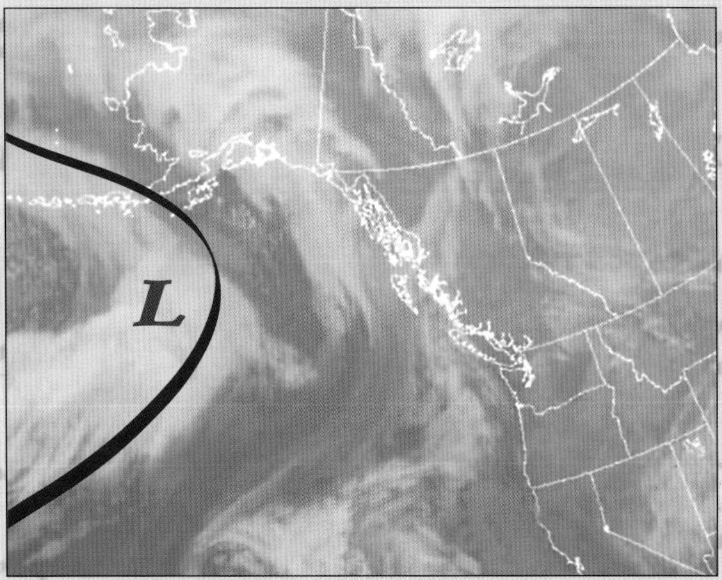

The wind diminishes but rain continues. A deep low pressure storm moves across the ocean.

Daytime Low Tide:	*9:14 am*	*3.0*
Daytime High Tide:	*3:06 pm*	*9.1*
Sunrise:	*7:09 am*	
Sunset:	*4:51 pm*	

By morning the Columbia's waters had flattened out smooth. Without hesitation the men loaded their canoes and set out.

> *Rained verry hard the greater part of last night and continues this morning. the wind has luled and the waves are not high; we loaded our canoes and proceeded on*
>
> Moulton, *Journals* 6:39 (Clark)

It must have been a tremendous relief to be moving forward once again. They were on their way, and in just another couple of hours the party would arrive at the Pacific Ocean.

As they paddled along the steep, forested shoreline, the men could easily see what lay ahead. This shore was actually a series of coves, each divided from the next by a small point of land that projected slightly out into the river. It was an ideal situation. Each cove provided a stretch of smooth, sheltered water. The men paddled close to shore, their canoes gliding along effortlessly.[1]

The weather calms; the river flattens. The men load and launch their canoes.

> *passed Several Small and deep nitch on the Stard. Side, we proceeded on about 10 miles*
>
> Moulton, *Journals* 6:39 (Clark)

Up ahead they could plainly see a prominent point sticking out from the shore. In fact, this dark, rocky headland extended out into the water

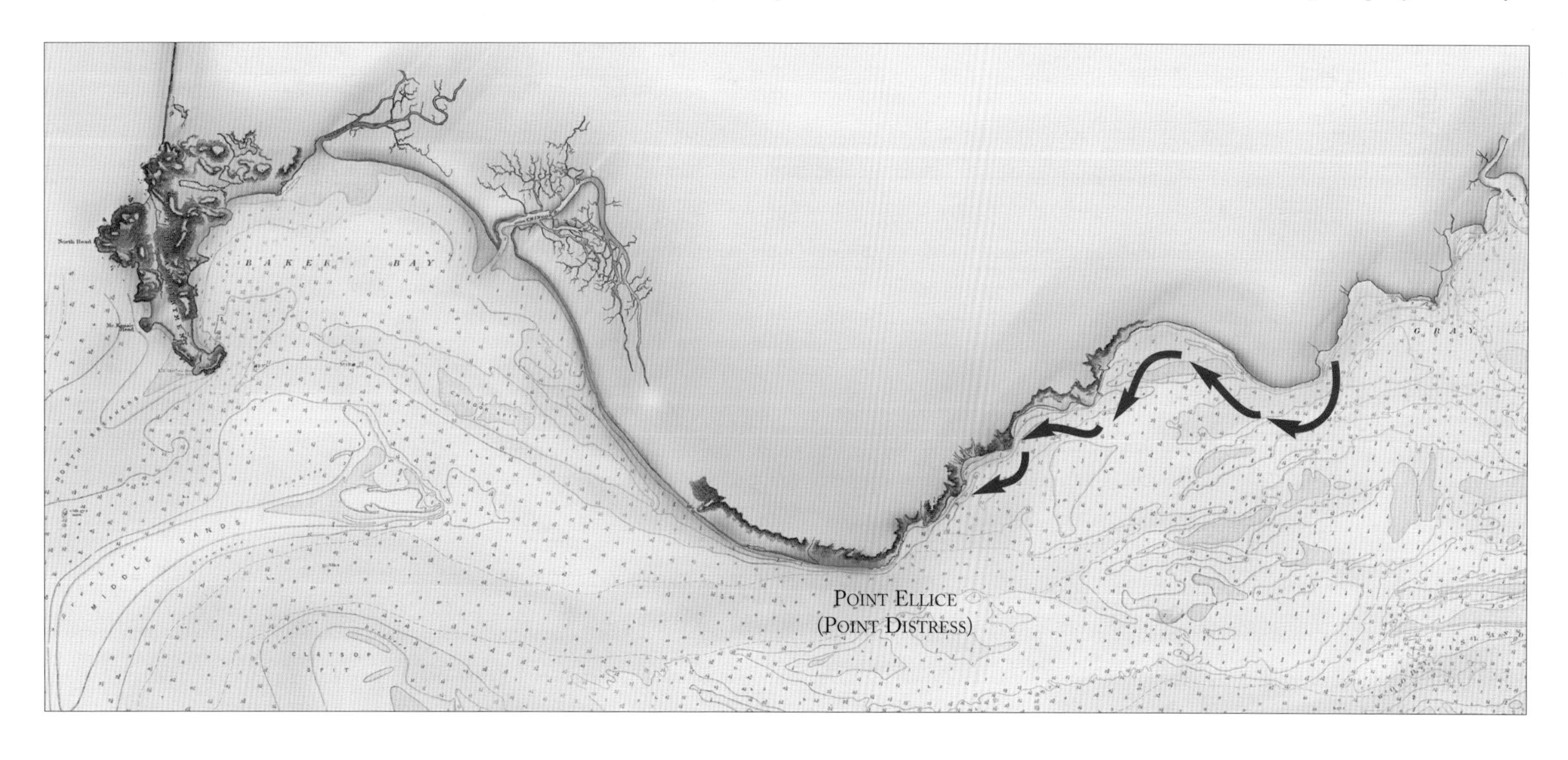

Lewis and Clark paddle downriver following close to the shore. The ocean is now only a couple of hours away.

(above) The men look downriver and see the last point of land between them and the ocean. Their journey west is nearly finished.

farther than any other and completely blocked their view downriver. It seemed clear that once they passed this point, they would be very, very close to the ocean.

Simultaneously, however, they were drawing nearer to the dreadful mouth of the Columbia. The wild and tumultuous waves from the Pacific Ocean were out of view, but the men undoubtedly began to feel the pulse of the powerful surf lifting and dropping the water beneath their canoes.

What occurred next is one of the most surprising moments of the entire expedition. As they drew nearer to the point, they saw a terrible and threatening sight. Waves pounded against the rocky shore, then swirled around, causing the river's current to boil into a whitewater chop. Huge driftwood logs floated among these waves, plunging below the surface then suddenly rising into view, like breaching whales. Water was surging at the party from every direction and slapping hard against their blunt canoes. In the distance, the roar of the ocean sounded like an enormous waterfall warning everyone that the worst was yet to come.[2]

Lewis and Clark could plainly see that if a canoe overturned, even their best swimmers would drown. Every impulse told them to keep going, to finish this journey; and yet all their common sense told them to wait. They had come too far to make a hasty, risky decision. In the end they decided to play it safe. Wisely, and for the first time in this expedition, Lewis and Clark ordered their party to turn around.

> *the waves became So high that we were compelled to return about 2 miles to a place we Could unload our Canoes, which we did in a Small nitch*
>
> Moulton, *Journals* 6:39 (Clark)

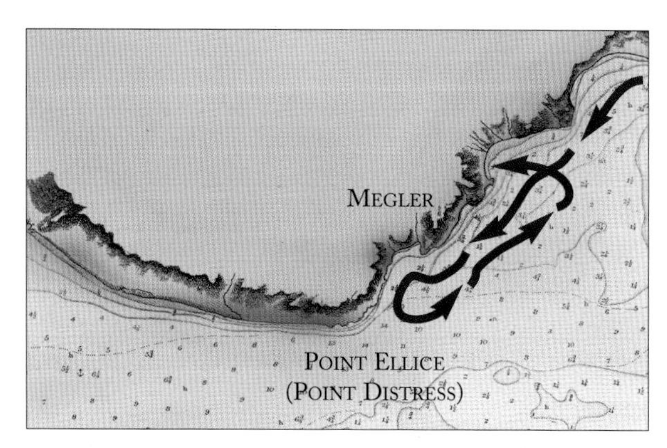

(right) As they approach the rocky point, Lewis and Clark find the waves dangerously high for their canoes. They turn around and beach their party in a nearby cove. Their first attempt around Point Distress has failed.

The men retreated upriver to a small cove they had passed moments earlier. They unloaded their canoes and built large fires. There was nothing to do but wait. The pouring rain underscored the misery of this unexpected setback.

Several hours later, however, Lewis and Clark noticed the river flattening again into a glossy, mirror-smooth surface. This was exactly the break they had been waiting for. The five canoes were loaded and launched, and the party set out downriver once again.

> *we continued on this drift wood untill about 3 oClock when the evening appearing favourable we loaded & Set out in hopes to turn the Point below and get into a better harber* Moulton, *Journals* 6:39 (Clark)

(left) In the afternoon the river appears calm. They load up their canoes and set out once again.

Unfortunately, those who are not familiar with tidewater can occasionally be fooled by its constantly changing conditions, and this is exactly what happened to Lewis and Clark. Their view upriver had revealed only a temporary calming of the waves, known as *highwater slack*. This natural phenomenon occurs when the flood tide collides with the ebb tide and the two opposing currents temporarily cancel each other. This illusion of calmness lasts only for a moment.[3]

As Lewis and Clark's party approached the point for the second time, they saw, to their surprise and horror, that the waters remained dangerous. Waves were rolling and breaking with great force and fury. The captains could plainly see that it would be absolutely impossible to paddle their canoes any further without risking everyone's lives, so once again they were forced to turn around.

> *but was obliged to return finding the waves too high for our Canoes to ride* Moulton, *Journals* 6:39 (Clark)

Now what should they do? Evening was upon them; darkness was less than two hours away. Obviously, they would have to find a campsite for the night, but where should they go? If they went too far upriver, and out of sight of this point, how would they know when the waters had become calm?

It made more sense to keep the party within sight of the point. Fortunately, there happened to be a tiny protected cove close at hand. A small spring trickled down the steep rocky hillside, supplying drinking water; firewood was everywhere. The site was exposed and barely adequate, but it would do for just one night, so they landed.

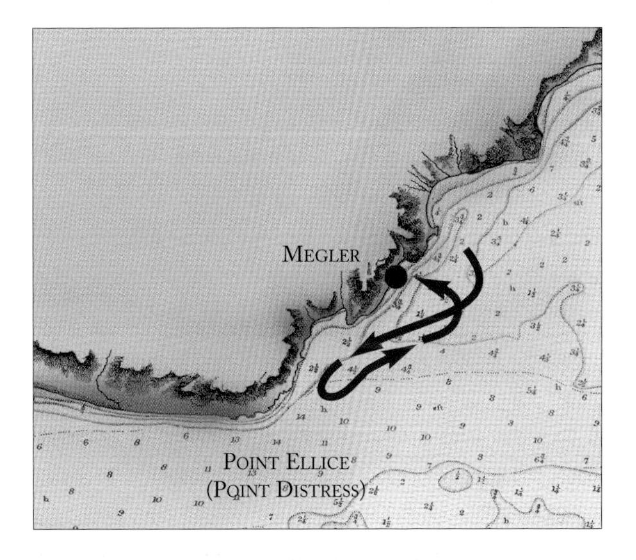

The waves are too high for their canoes. Lewis and Clark turn around once again, and land in a nearby cove.

> *we again unloaded the Canoes, and Stoed the loading on a rock above the tide water, and formed a camp on the Drift Logs which appeared to be the only Situation we could find* Moulton, *Journals* 6:39 (Clark)

The ocean surge breaks violently against Point Distress.

They rolled away large rocks, set the driftwood on fire for warmth, and wrapped themselves up against the cold steady rain. It was going to be another long wet night, but with luck they could set out early the next morning.

> *The rain Continud all day – we are all wet, also our beding and many other articles. we are all employed untill late drying our bedding. nothing to eate but Pounded fish* Moulton, *Journals* 6:39 (Clark)

The turbulent, unpredictable waters of the mouth of the Columbia River had brought the Lewis and Clark expedition to a complete standstill. At this moment they probably thought it was a little setback, but soon they would discover that the next several days would be among the most grueling and depressing ordeals of their entire expedition.

Winter Storms

The Pacific Northwest is frequently hit by fierce winter gales between the months of October and April. These storms originate far out in the Pacific Ocean where they absorb moisture and move eastward, putting them on a collision course with the coast.

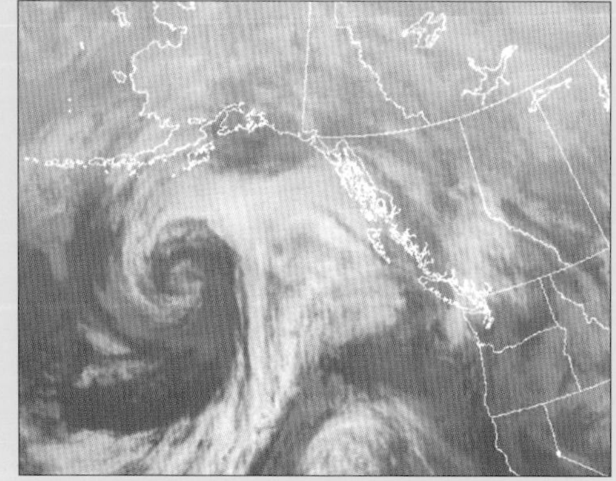

As the storm approaches the coastline, the wind typically switches to the southwest and blows with a sustained speed of thirty to forty miles per hour, accompanied by a heavy, relentless rain. Powerful gusts often exceed ninety miles per hour; rain is driven sideways in great horizontal sheets, and large trees tip over.

Some storms diminish within four hours; others pound with full fury for more than an entire day. Occasionally two or three storms stack up, resulting in three or four days of constant howling wind and torrents of rain. However, typically a break of fair, dry weather follows each gale.

Indian Nations

Lewis and Clark encountered various "nations" of Indians along the Lower Columbia. The Chinook and Clatsop resided near the ocean on opposite sides of the river; the Wahkiakum and Cathlamet occupied villages farther upriver.[1]

The four "nations" as described by the captains shared many similarities. Their houses, clothes, and basketry appeared practically identical; their language was indistinguishable.

In addition to the resident Indians, Lewis and Clark were visited by distant nations.

> *Several Indians Visit us to day of different nations or Bands Some of the Chiltz Nation who reside on the Sea Coast near Point Lewis, Several of the Clotsops who reside on the opposite Side of the Columbia immediately opposite to us, and a Chief from the Grand rapid to whome we gave a Medal*[2]

The coastal Indians lived in a rich environment where food was frequently gathered while the tide was low. Clark observed the following:

> *I Saw Indians walking up and down the beech which I did not a first understand the Cause of, one man came to where I was and told me that he was in Serch of fish which is frequently thrown up on Shore and left by the tide, and told me the "Sturgion was verry good" and that the water when it retired left fish which they eate*[3]

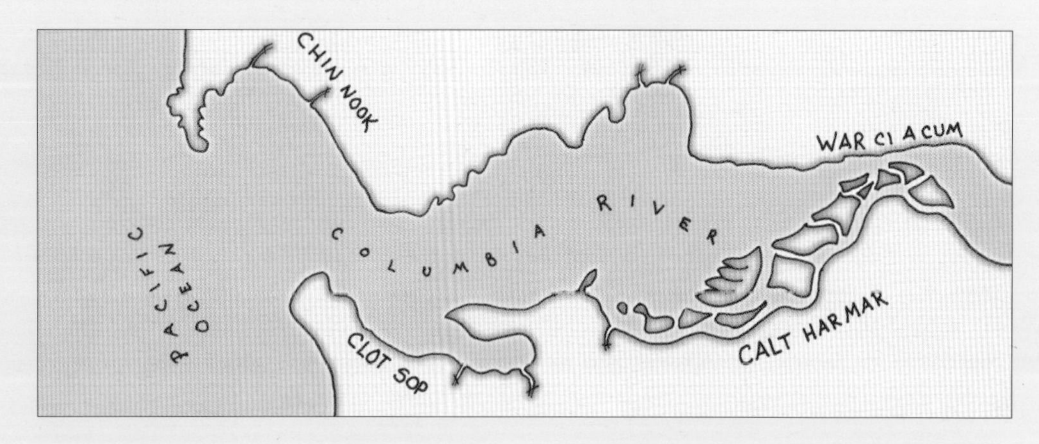

The Clatsops were fond of gambling. Clark described their favorite amusement in this manner:

> *those people have a Singular game which they are verry fond of and is performed with Something about the Size of a large been [bean] which they pass from, one hand into the other with great dexterity during which time they Sing, and occasionally, hold out their hands for those who Chuse to risque their property to guess which hand the been is in*[4]

Lewis and Clark noted that coastal Indian men and women held nearly equal positions of authority and shared in the labor. Lewis writes:

> *in common with other savage nations they make their women perform every species of domestic drudgery. But in almost every species of this drudgery the men also participate . . . nothwithstanding the survile manner in which they treat their women they pay much more rispect to their judgement and opinions in many respects than most Indians nations; their women are permitted to speak freely before them, and sometimes appear to command with a tone of authority: they generally consult them in their traffic and act in conformity to their opinions*[5]

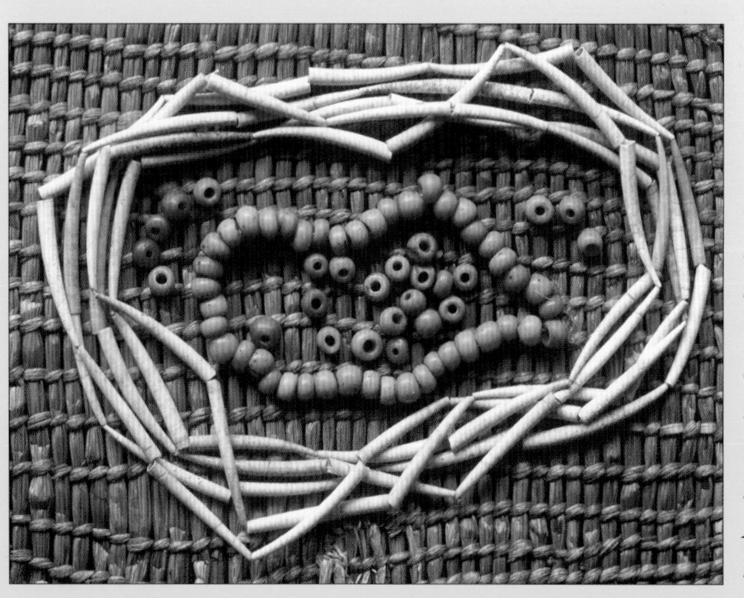

Rocks were used as weights for fishing lines and nets; larger rocks could anchor canoes.

Long slender sea shells called "dentalium" were traded among Indians and used as ornaments for their hair and clothes. After Euro-American contact, blue beads became equally popular.

Monday, November 11th

Rain continues throughout the morning, increasing in the afternoon. Cold air moves down from the north as a deep low pressure storm approaches the coast.

Daytime Low Tide:	*10:15 am*	*3.2*
Daytime High Tide:	*4:01 pm*	*8.6*
Sunrise:	*7:10 am*	
Sunset:	*4:51 pm*	

IN THE PACIFIC NORTHWEST THE WINTER nights are fourteen hours long. What made these nights seem even longer for Lewis and Clark's men was the fact that their tents had completely rotted. Perhaps the captains themselves slept beneath some sort of awning, but most of the men simply curled up like wild beasts, trying to ignore the cold torrents of rain. Private Joseph Whitehouse describes this unbelievably dismal night:

> *It rained hard the greater part of last night, which made it very disagreeable to us all. The greater part of our Men had nothing to shelter them from the rain, & were obliged to lay down in it, & their Cloathes were wet through*
>
> Moulton, *Journals* 11:392 (Whitehouse)

Daylight revealed their worst fears. The waves in the river were too rough for their canoes.

> *This morning continued wet & rainey, the wind was high, & the swell in the river ran very high, & We did not attempt to move from this place*
>
> Moulton, *Journals* 11:392 (Whitehouse)

Three days had passed since the men had eaten their last meal of fresh meat, so Clark sent hunters into the surrounding hills to shoot deer, elk, bear, or anything else they could find. However, after a brief absence, these hunters returned to camp empty-handed. The thick coastal rainforest, which requires as much climbing as it does hiking, had turned them back. Clark describes their discouraging return:

> *Sent out Jo Fields to hunt, he Soon returned and informed us that the hills was So high & Steep, & thick with undergroth and fallen Timber that he could not get out any distance*
>
> Moulton, *Journals* 6:41 (Clark)

It appeared as though this would be another day of hunger, but luckily a canoe full of Indian merchants came paddling downriver, and they happened to have a load of fish on board. The Indians landed, and the captains immediately seized upon this opportunity to purchase a fresh meal for their men.

Waves continue around Point Distress. Lewis and Clark are unable to launch their canoes.

"the hills was So high & Steep, & thick with undergrowth and fallen Timber"

The fallen timber Joseph Field described to Clark was quite unlike any obstacle these men had ever encountered. They found the woods crisscrossed with gigantic trees measuring eight to twelve feet in diameter that had fallen as a result of the high winds or old age. Every windfall created a natural barricade. Dressed as he was in wet leather and carrying a rifle in one hand, it would have been exhausting for Joseph Field to constantly climb over the massive trees, and the commotion would have frightened any game.

The logger pictured here with his saw, circa 1905, serves to show the enormous scale of this fallen timber.

> *at 12 oClock at a time the wind was verry high and waves tremendeous five Indians Came down in a Canoe loaded with fish . . . we purchased of those Indians 13 of these fish* Moulton, *Journals* 6:40 (Clark)

This fish[4] was a welcome treat for everyone, but even more exciting was the news of where the Indians had intended to take their catch.

> *they are on their way to trade those fish with white people which they make Signs live below round a point* Moulton, *Journals* 6:40 (Clark)

This was exciting information: the party had heard rumor after rumor about fur traders living near the ocean, but here was proof. According to these Indians, there were white people less than a mile away.

Lewis and Clark knew that as soon as they made contact with fur traders, most of their troubles would be at an end. Traders would have a variety of supplies on hand that could re-outfit the entire party with new tents, clothes, food, tobacco, and maybe even a dram of whiskey. They were certainly lucky these Indians had stopped by their camp.

Having sold their cargo of fish, the Indians set out to return to their village. However, instead of hugging the shoreline, where the waves were somewhat calm, they paddled straight out into the middle of the vast Columbia and proceeded to cross over to the opposite side through the most horrible waves into which any of these men had ever seen anyone take a canoe.

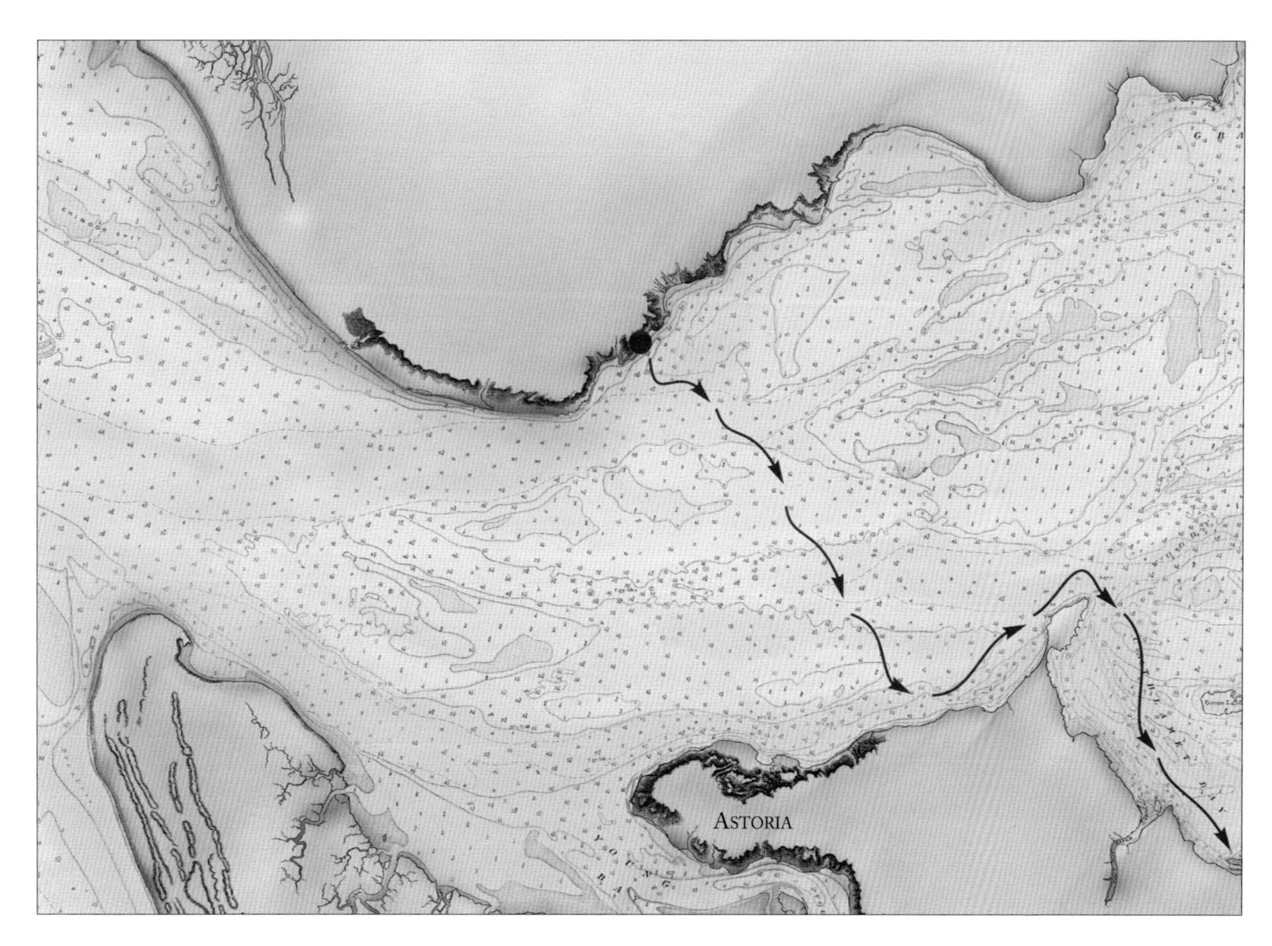

The Indian merchants cross the Columbia and return home.

Elegant Canoes

The Indians of the Northwest Coast built the most perfect watercraft Lewis and Clark had ever seen. While in the lower Columbia River the captains noticed five different classes of canoes; some were possibly built by the Chinook or Clatsop people, others may have been sold to the local people by tribes residing in the North. Lewis writes:

> *The Canoes of the natives inhabiting the lower portion of the Columbia River make their canoes remarkably neat light and well addapted for riding high waves*[1]

These canoes were carved from single logs, then softened with hot water and steam until they could be bent to achieve a precisely shaped hull that allowed them to slice through waves of any size.[2] In addition to this, the natives were expert canoeists.

> *I have seen the natives near the coast riding waves in these canoes with safety and apparently without concern where I should have thought it impossible for any vessel of the same size to lived a minute*[3] (Lewis)

The most common canoe measured thirty-five feet in length;[4] however, smaller ones, less than half that length, were used for hunting and gathering. The largest canoes were true masterpieces of engineering.

> *some of the large canoes are upwards of 50 feet long and will carry from 8 to 10 thousand lbs. or from 20 to thirty persons and some of them particularly on the sea coast are waxed painted and ornimented with curious images at bough and Stern; those images sometimes rise to the hight of five feet*[5] (Lewis)

Makah canoe carver, 1914

Lummi fisherman, c. 1900

Clark described how canoes were made. "The only tool usually employd in forming the Canoe, carveing &c is a chissel formed of an old file about an inch or 1½ inchs broad . . . a person would Suppose that forming a large Canoe with an enstriment like this was the work of Several years; but those people make them in a fiew weeks."[6]

Canoes had to negotiate a wide variety of water conditions. Some were designed for use in the open ocean, others remained in rivers and bays. At first glance these canoes appear similar; however, styles varied from nation to nation and no two canoes are identical. Considerable expertise is required to identify subtle differences particular to a given region.

Makah whalers, c. 1900

> *the Indians left us and Crossed the river which is about 5 miles wide through the highest Sees I ever Saw a Small vestle ride*
>
> Moulton, *Journals* 6:40 (Clark)

Lewis and Clark had admired the elegant canoes the Indians made, but until this moment they had never seen them used in challenging waters by the natives themselves. What they now saw exceeded every expectation. Instead of rolling over and capsizing, the Indians sliced easily through the whitecapped waves.

> *their Canoe is Small, maney times they were out of Sight before the were 2 miles off*
>
> Moulton, *Journals* 6:40 (Clark)

The men were dumbstruck. They had never imagined such a feat possible. This remarkable display of expertise drew words of great admiration and praise from Captain Clark.

> *Certain it is they are the best canoe navigators I ever Saw* Moulton, *Journals* 6:40 (Clark)

As the afternoon wore on, the rain increased to a steady downpour. Soon the saturated hillside had softened, loosening the rocks and sending them tumbling directly into the party's camp. Rocks flew past the men, ricocheting off the driftwood with the force of cannonballs. The pumpkin-sized ones were easy to avoid, but the smaller rocks, about the size of an apple, were a greater problem. It was practically impossible to see them coming, and they hit with enough force to break bones.

> *the great quantities of rain which has fallen losenes the Stones on the Side of the hill & the Small ones fall on us* Moulton, *Journals* 6:40 (Clark)

Danger was all around. It was too risky to keep everyone together, so the men split up and crawled into any small space they could find to avoid the falling rocks. There would be no campfire tonight. As darkness approached, it was every man for himself.

> *our Selves & party Scattered on drift trees of emense Sizes, & are on what dry land they can find in the Crevices of the rocks & hill Sides*
>
> Moulton, *Journals* 6:40 (Clark)

Their situation was becoming desperate, yet there was nothing Lewis and Clark's party could do. Their only choice was to hunker down to keep themselves safe.

Tuesday, November 12th

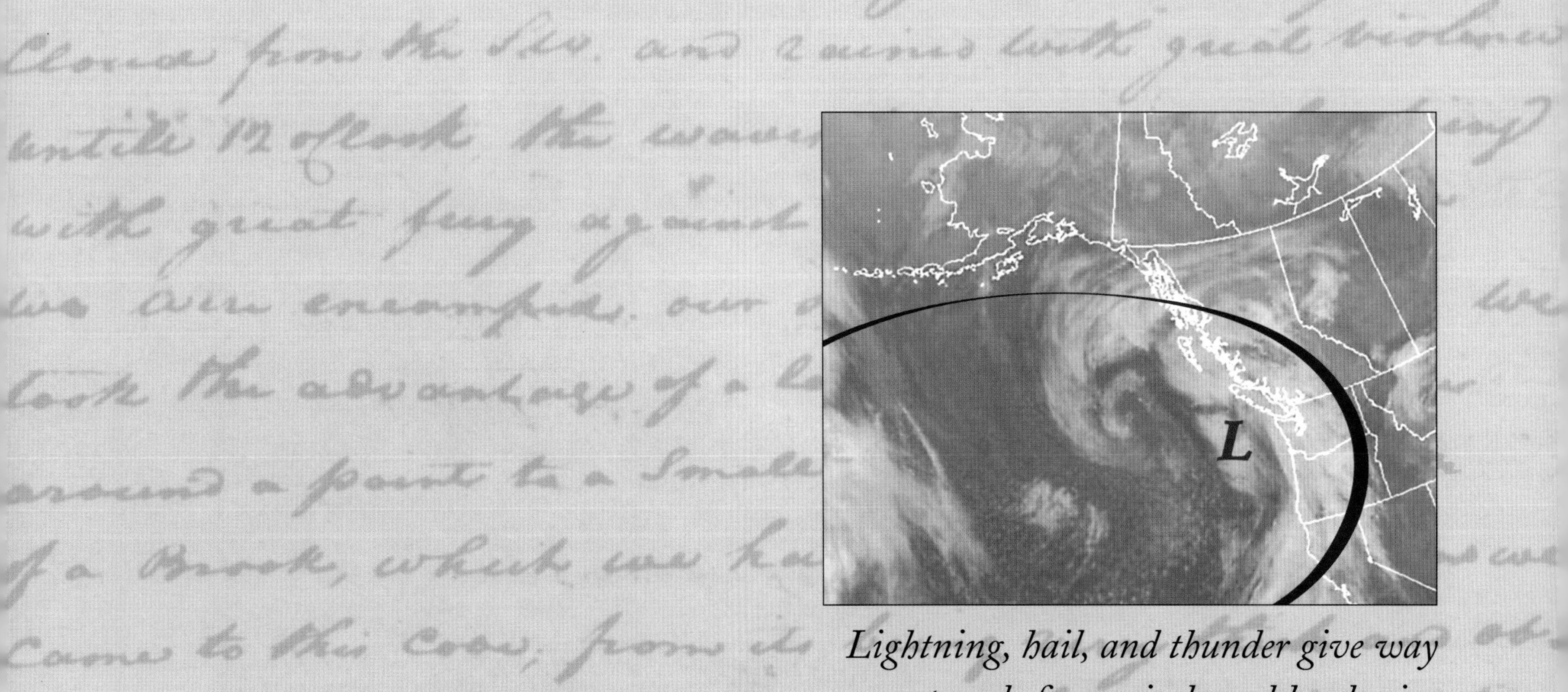

Lightning, hail, and thunder give way to gale force winds and hard rain.

Daytime Low Tide:	*11:25 am*	*3.2*
Daytime High Tide:	*5:06 pm*	*8.0*
Sunrise:	*7:12 am*	
Sunset:	*4:50 pm*	

The good luck that Lewis and Clark had enjoyed throughout most of their expedition seemed to have vanished. Everything was going wrong; every day their predicament became gloomier than the day before.

Just when they most needed a break, their situation took another turn for the worse. In the middle of the night, they were awakened by a rare thunderstorm. The sky rumbled with thunder, as lightning flashed and buckshot-sized hailstones pelted the earth. When hail falls on the wide Columbia the river hisses like high pressure steam.

> *A Tremendious wind from the S.W. about 3 oClock this morning with Lightineng and hard claps of Thunder, and Hail which Continued untill 6 oClock a.m.* Moulton, *Journals* 6:43 (Clark)

The hailstones stung like wasps and brought a terrible chill into the air, but the men didn't dare move around. The falling rocks were far more dangerous than the shivering cold temperatures. After several hours of this agonizing predicament, the storm passed, dawn came, and the sky brightened. The worst seemed to be over, but what looked like the beginning of good weather was merely the calm before an even worse storm.

> *it became light for a Short time, then the heavens became Sudenly darkened by a black Cloud from the S.W. and rained with great violence*
>
> Moulton, *Journals* 6:43 (Clark)

Now, for the first time, the Lewis and Clark party experienced a full gale directly from the ocean. The wind drove the rain sideways and pushed the waves into the shore with such force

The waves increase in size and force.

that the spray showered down upon them. At any moment the force of this water could roll a driftwood log and crush them to death; or worse yet, a huge wave might sweep them into the water where they would perish in seconds.

> *the waves tremendious brakeing with great fury against the rocks and trees on which we were encamped. our Situation is dangerous*
>
> Moulton, *Journals* 6:43 (Clark)

If they backed up from the waves, they were beneath the falling rocks; if they moved away from the steep hillside, they were within range of the waves. They were vulnerable on every side, and there was absolutely no place to go.

It was clear that they had to abandon this camp, but where would they go and how would they get there? It would be impossible to launch their canoes. These huge waves would flip them over in ten seconds and drown the entire party. Behind them the hills were so steep and rocky that even their best hunters, carrying only a rifle, couldn't find a path. There seemed to be nothing to do but wait.

Then, at what seemed to be the darkest moment, their luck returned. The tide was ebbing, and even despite the hard wind blowing in from the ocean, the level of the river dropped several feet. A narrow, rocky toe of shoreline emerged at the bottom of the steep cliffs, and the captains instantly saw it as a possible way out of this situation. If the water level kept dropping, they could abandon this camp, dash on foot along the edge of the shore and perhaps find refuge in another cove.

This was a fine plan except for one problem. If they abandoned this camp, what could they do with their canoes? Several days earlier, the canoes had nearly been crushed between the driftwood logs, and the same thing could happen again if they were left unattended. Yet there was no way the men could drag these huge, heavy canoes high enough up the steep bank to be out of harm's way.

There seemed to be only one solution, and even it was risky. They would bury their canoes: bury them beneath tons of rock. If each canoe were weighted securely to the bottom of the river, perhaps the waves and driftwood would pass harmlessly over the top, leaving them safe.

Large stones were lying everywhere, so it would have taken Lewis and Clark's men only a couple of minutes to fill each canoe full of rock. However, their thin cedar Indian canoe posed a problem. This delicate craft would probably split down the middle if filled with rock, so the captains decided it should not be left behind. It was so light that four men could easily carry it; besides, keeping some means of transportation with them was a very good idea.[5]

Now the only remaining concern was their baggage. Since there was too much of it to carry, the men selected only the bare essentials, such as axes, blankets, and kettles. The rest was stowed high up on the hillside above the reach of the crashing waves. They could come back later for their medicines, blacksmith tools, survey instruments, and trading goods.

The party watched and waited. When at last the ebbing tide seemed to reach its lowest point, they set out.

> *we took the advantage of a low tide and moved our camp around a point to a Small wet bottom at the mouth of a Brook* Moulton, *Journals* 6:43 (Clark)

The captains led the party along the narrow shore with waves crashing all around. The force of the water occasionally knocked them to their knees. The wind howled, and thousands of chunks of driftwood thumped ceaselessly against the rocks, making the sound of horses galloping across a wooden bridge. They struggled along the slippery rocks, past a steep cliff, and stepped into the mouth of a small stream. The entire move would have taken less than fifteen minutes.

Suddenly they found themselves walking into a narrow, dim canyon between two hillsides of enormous, dark trees. The air felt as cold as ice. The ground was saturated from the constant rainfall; moisture squished beneath every step like a wet sponge. Overhead the canopy of gigantic cedar and spruce trees was so thick that sunlight rarely reached the ground. This was their first contact with Northwest rainforest, and it was no doubt the dampest, darkest, most dismal forest any of these men had ever seen.

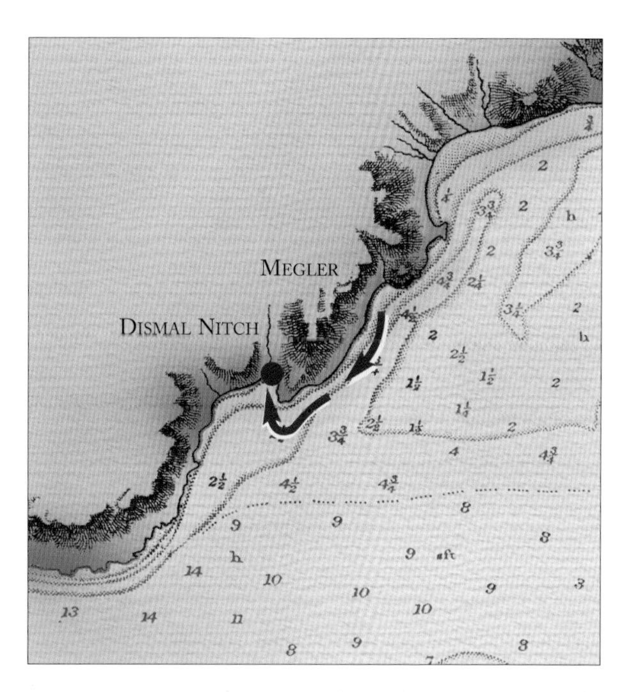

Lewis and Clark bury their canoes, abandon their baggage and set out on foot. They proceed along the shore, enter a narrow canyon, and set up camp in Dismal Nitch.

(below) Coastal rainforest.

Spruc Pine grow here to an emense Size & hight maney of them 7 & 8 feet through and upwards of 200 feet high Moulton, *Journals* 6:42 (Clark)

As inhospitable as it might have first appeared, there was indeed plenty of room to camp in this narrow little canyon. Here there was no immediate danger. Nothing was tumbling down the hillsides, and the crashing waves could not reach into this ravine.

Food was their first priority, so the captains sent out hunters to shoot deer or elk, but they soon returned with disappointing news and complaints about the rugged, inaccessible woods. In the meantime, Clark and his men chased spawning salmon up and down the shallow creek. The meat from these overly-mature fish might have been soft and bland, but certainly better to eat than the dried Indian fish they carried in reserve.[6]

Send out men to hunt they found the woods So thick with Pine & decay timber and under groth that they could not get through . . . I walked up this creek & killed 2 Salmon trout, the men killd. 13 of the Salmon Species Moulton, *Journals* 6:42 (Clark)

Salmon in the lower Columbia River are bright silver, handsome fish, but soon after entering rivers to spawn their flesh softens, they turn dark in color, and their jaws curve open, exposing large canine teeth.

Hail from the night before remained scattered on the ground, which meant that the temperature of the air could not have been much above thirty-eight degrees. These men were not strangers to snow or frost, but this cold, wet coastal climate affected them differently. The icy-cold rain trickled down the men's faces and dripped from their soaking wet clothing, bringing on a chill that penetrated deep into their joints. Old injuries, bumps, and bruises throbbed. Their knuckles swelled, causing their fingers to curl up into useless fists. They needed warming fires, but holding an axe handle tight enough to cut was practically impossible. Besides, every chunk of wood was so wet that it hissed and smoldered, then eventually went out.[7]

Captain Clark looked at his men. He saw the uncontrollable shivering, the chattering teeth, and the grimace on every sunken face. He realized they had reached a new low point. The men had suffered before, but this situation was more hopeless and pitiful than anything they had ever experienced. What would Thomas Jefferson or his colleagues at the American Philosophical Society say if they could see the suffering of these men? Clark wrote in his journal:

It would be distressing to a feeling person to See our Situation at this time all wet and cold with our bedding &c. also wet, in a Cove Scercely large nough to Contain us Moulton, *Journals* 6:42 (Clark)

Never before had Lewis and Clark been forced to abandon and bury their canoes. Never before had they walked away from their precious supplies, so Clark carefully documented this exceptional circumstance.

our Baggage in a Small holler about 1/2 a mile from us, and Canoes at the mercy of the waves & drift wood, we have Scured them as well as it is possible by Sinking and wateing them down with Stones to prevent the emence waves dashing them to pices against the rocks Moulton, *Journals* 6:42 (Clark)

This was a setback, but the captains were not giving up. They immediately laid out a plan to get a few members of the party downriver. Their small canoe was the same style used by the Indians who had brought the fish. It seemed logical that if the Indians could navigate through high waves, then so could they. Three men were selected to take this canoe downriver to search for the white men below.

Three men Gibson Bratten & Willard attempted to decend in a Canoe built in the Indian fashion and abt. the Size of the one the Indians visited us in yesterday Moulton, *Journals* 6:42-43 (Clark)

They set out with high hopes of success, but within moments the waves were slapping them around in circles. The swells were too high, the current too strong. This third attempt to get around Point Distress failed miserably, and the three men were lucky to return alive.

they proceeded to the point from which they were oblige to return, the waves tossing them about at will Moulton, *Journals* 6:43 (Clark)

In a desperate attempt to make contact with fur traders, a canoe is sent out. The waves at Point Distress force the crew to return. This third attempt fails.

Having no other choice, the party wrapped themselves in their blankets as best they could and prepared for nightfall. They ate their dinner of spawned salmon, then found a place to lie down. Water wicked up from beneath them as the rain showered down from overhead. During such misery, time has no meaning. What seems like an hour might actually be only ten minutes. These days were probably beginning to blend into each other. How long had they been stuck here? What day was it? How many days had passed since they first saw the ocean?[8] Clark was right. It would indeed have been distressing to a compassionate person to see what suffering these men were forced to endure during the final miles of their westward journey.

Wednesday, November 13th

The storm moves northward as the jet stream pushes up from the south. Rain and wind continue along the coast.

Daytime High Tide:	*6:48 am*	*7.3*
Daytime Low Tide:	*12:40 pm*	*2.9*
Sunrise:	*7:12 am*	
Sunset:	*4:49 pm*	

The Columbia River did not grow calm during the night. The next morning the captains saw waves breaking from shore to shore and knew that it would be impossible to launch their canoes in such dreadful conditions.

Instead of wasting the day waiting for the weather to change, Clark turned his back to the river and set out. He planned to hike up to a hilltop for a better view of the lower river.

> *I walked up the Brook & assended the first Spur of the mountain with much fatigue, the distance about 3 miles, through an intolerable thickets*
>
> Moulton, *Journals* 6:44 (Clark)

This was Clark's first attempt to walk deep into the rainforest, and it did not go well. He became entangled in the brush and was slowed down when he had to pull himself, hand over hand, up the steep hills.

> *arrow wood with briers, growing to 10 & 15 feet high interlocking with each other & Furn, added to this difficulty the hill was So Steep that I was obliged to drawing my Self up in many places by the bowers*
>
> Moulton, *Journals* 6:44 (Clark)

Lewis remained in camp while Clark explored in the hills. We have no description of how he passed the time, but perhaps he and his men used this opportunity to portage their supplies into the new camp and pull each canoe, one by one, along the shore. It made perfect sense to keep all their baggage and canoes and men together in the same camp. However, since the journals don't mention this it remains speculative.

Clark reached the top of a hill but couldn't see anything. His trip ended up being a complete waste of time.

> *my principal object in assending this mountain was to view the countrey below, the rain continuing and weather proved So Cloudy that I could not See any distance*
>
> Moulton, *Journals* 6:45 (Clark)

(left) Clark follows the stream up into the hills, then ascends a hilltop for a view downriver.

(right) Lewis remains in camp. The men have an opportunity to retrieve their abandoned canoes and baggage and move them into Dismal Nitch.

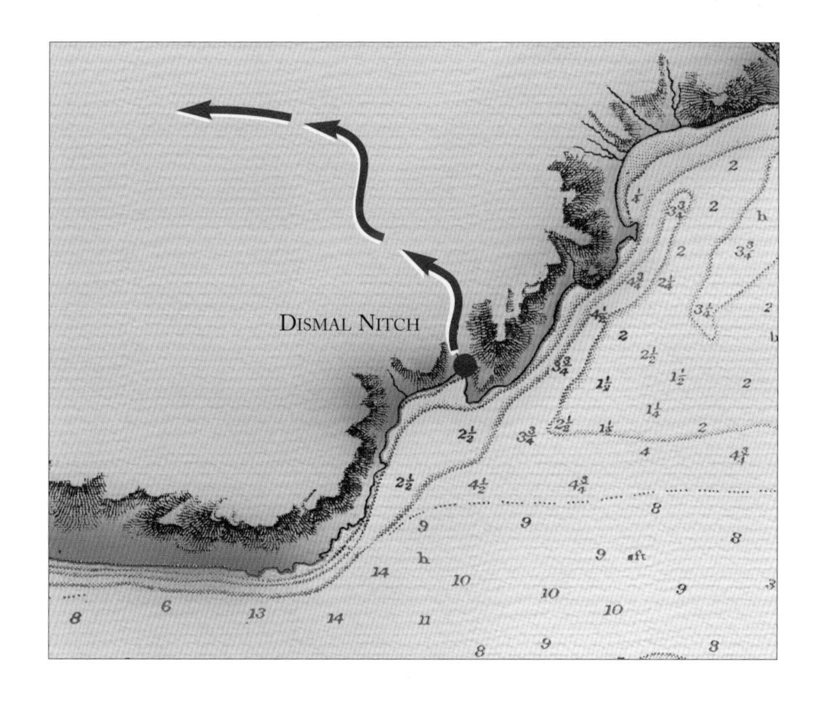

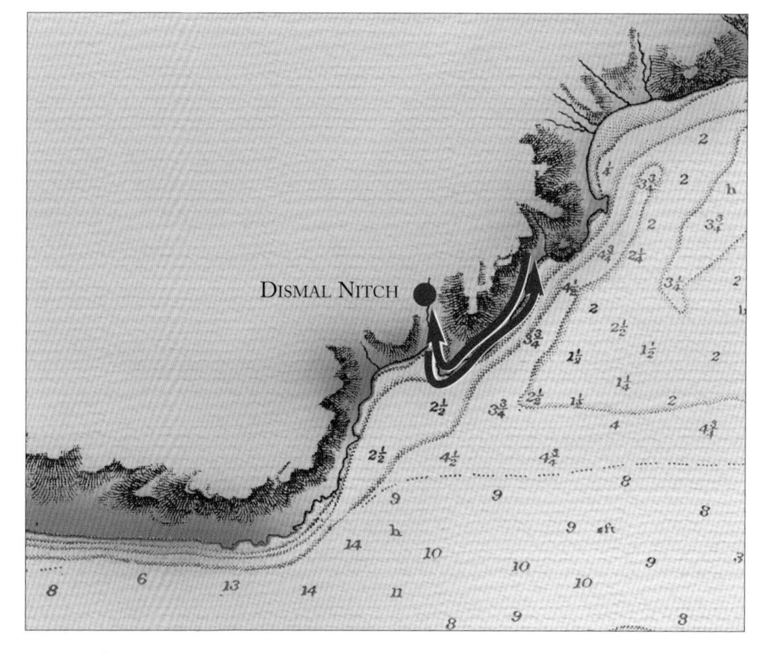

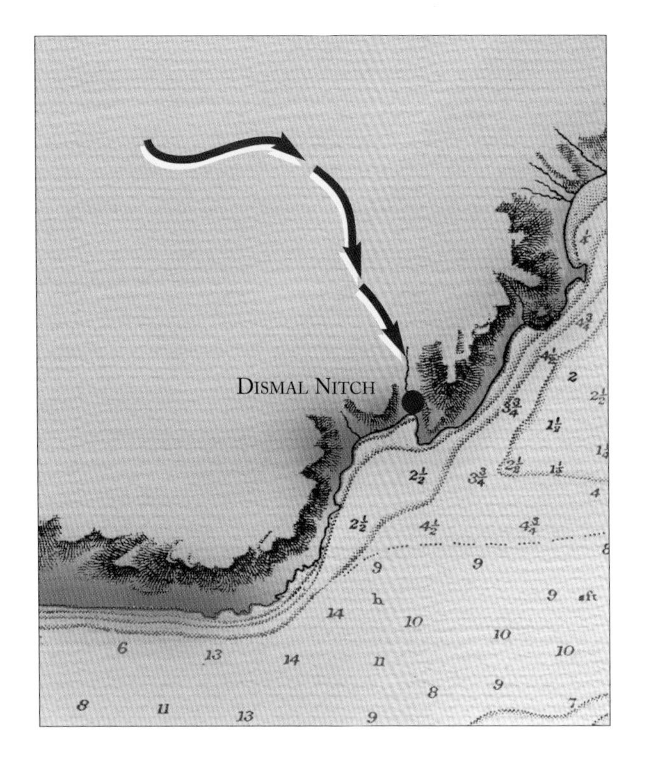

Clark returns; his effort to get a glimpse of the lower river has failed.

Clark had now experienced what his hunters had been complaining about; it would be impossible for the party to travel on foot. Those steep, rugged hills were practically impenetrable. The captains might not have wanted to admit it, but both men knew that they had gotten themselves into a terrible predicament.

Making contact with the fur traders who lived around the point seemed to be their only hope of getting out of this miserable cove, so despite the three previous failures the captains determined to try once again. Their indispensable Indian canoe was set into the water and another crew assembled. Alexander Willard had failed on the first attempt the previous day but was willing to try again. John Colter and George Shannon rounded out the crew.

The waves remain high; the cold waters of the Columbia churn and roll violently against the black, rocky point.

> *on my return we dispatched 3 men Colter, Willard and Shannon in the Indian canoe to get around the point if possible and examine the river, and the Bay below* Moulton, *Journals* 6:45 (Clark)

This was going to be a risky move, but they had no other choice. If there was someone living on the other side of this point, perhaps these men could make contact and return before dark. Despite the rough water, the captains thought it was worth a try. Clark described the condition of the river.

> *The Tides at every flud come in with great Swells & Breake against the rocks & Drift trees with great fury* Moulton, *Journals* 6:44 (Clark)

The canoe was launched, and the crewmen paddled hard into the treacherous waves. Soon they were out of sight. The party waited and waited for the three men to return. Evening came, then darkness.

> *our canoe and the three men did not return this evening* Moulton, *Journals* 6:44 (Clark)

Lewis and Clark had no idea what had became of the three men. It was entirely possible that they had found a camp of fur traders and were assembling a rescue party to return early the next morning. However, it was equally possible that the waves had driven them into the rocks, that they had capsized and drowned. This uncertainty about their fate would have made the nighttime feel blacker, colder, and longer than ever before.

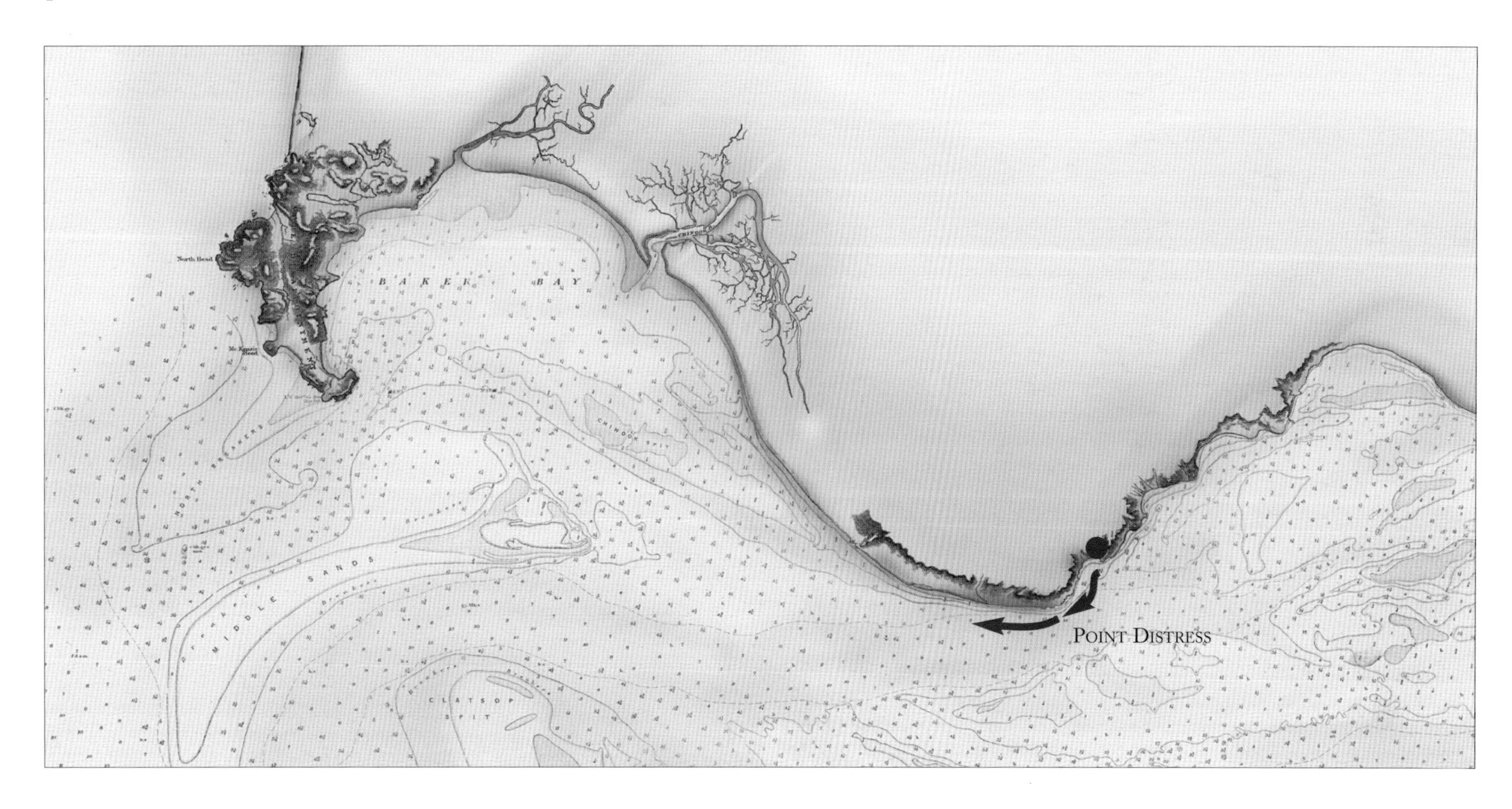

Willard, Shannon, and Colter set out from Dismal Nitch in their Indian canoe. This is the party's fourth attempt around Point Ellice (Point Distress).

Thursday, November 14th

Rain continues throughout the day but diminishes as the jet stream moves over central Oregon.

Daytime High Tide:	*7:47 am*	*7.6*
Daytime Low Tide:	*1:56 pm*	*2.4*
Sunrise:	*7:14 am*	
Sunset:	*4:49 pm*	

THE MORNING LIGHT REVEALED AN unchanged Columbia River. It churned and heaved its waters in a most threatening manner. The *Graveyard of the Pacific* was defending its well-earned reputation.

> *Rained last night without intermission and this morning the wind blew hard . . . We Could not move* Moulton, *Journals* 6:46 (Clark)

In addition to this, during the night one of the party's large canoes had been severely damaged. Now they couldn't leave here even if the river flattened out calm.

> *one of our Canoes is much broken by the waves dashing it against the rocks* Moulton, *Journals* 6:46 (Clark)

(left) The waters of the Columbia remain unchanged.

Three men were missing without a trace. The seaworthy Indian canoe was gone, and now one of their large canoes had cracked open. Little by little the expedition seemed to be falling apart; something had to be done immediately to reverse this situation, so Lewis stepped forward with a plan.

Realizing it was useless and foolish to risk another canoe on a fifth attempt around Point Distress, Lewis announced his intention to hike overland down to the ocean. Their only hope at this moment was for him to make contact and get help from whoever was there.

> *Capt Lewis concluded to proceed on by land & find if possible the white people the Indians Say is below*
>
> Moulton, *Journals* 6:47 (Clark)

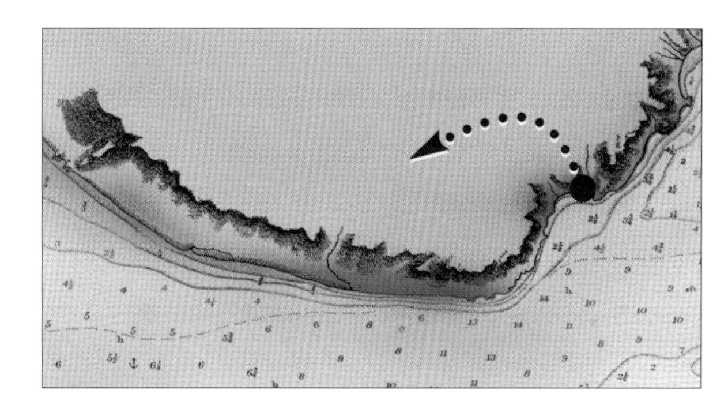

Lewis plans to hike overland and find the fur traders rumored to live near the ocean.

While Lewis began to prepare for this challenging hike through the rugged rainforest, several men applied their carpentry skills to repairing the broken canoe. During their journey down the Columbia, these men had become quite expert at such repairs and knew exactly what to do.[9] Suddenly their work was interrupted when an Indian canoe was sighted coming upriver around the point.

> *5 Indians Came up in a Canoe, thro' the waves, which is verry high and role with great fury*
>
> Moulton, *Journals* 6:46 (Clark)

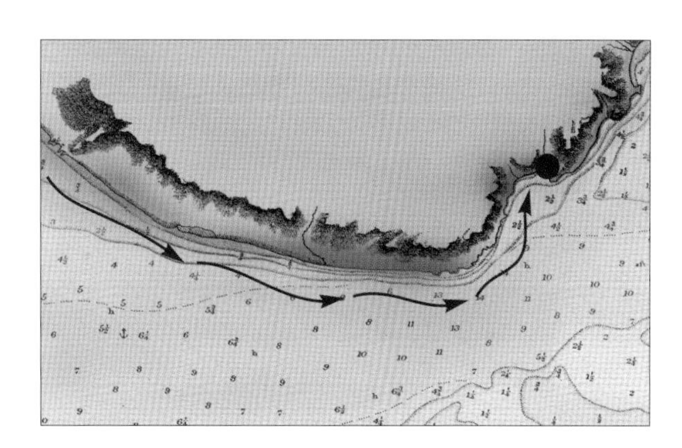

Three Wahkiakum men and two women stop at Dismal Nitch on their way upriver.

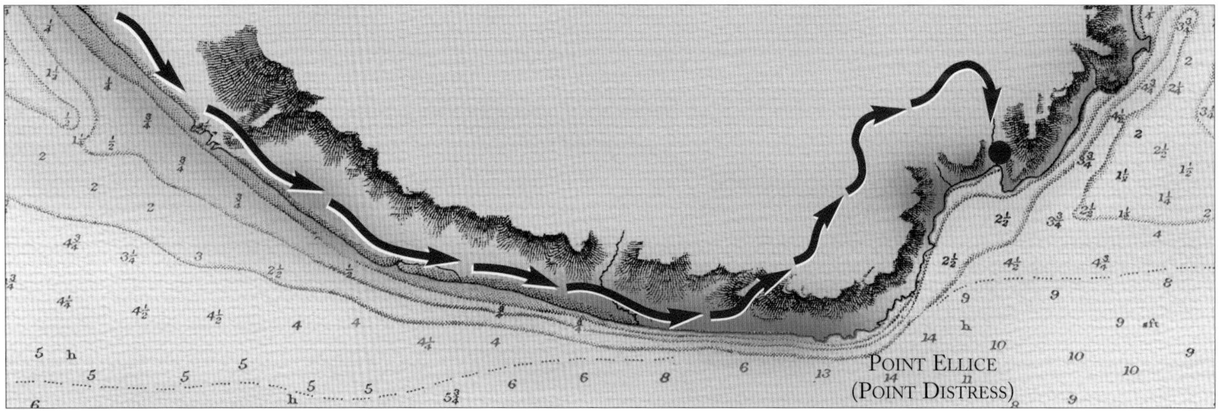

POINT ELLICE
(POINT DISTRESS)

John Colter hurries along the shore, climbs over Point Distress, and arrives back in Dismal Nitch.

No one, at this moment, could have imagined what an unlikely series of events was about to unfold. The three Indian men landed, while their two women paddled the canoe out into deeper water, apparently to keep it away from the shallow, rocky shore. Unable to speak English, the Indians used hand gestures to communicate with the party.

> *They made Signs to us that they Saw the 3 men we Sent down yesterday* Moulton, *Journals* 6:46 (Clark)

This was excellent news. Everyone's spirits must have been lifted upon hearing that the three men had survived the journey downriver. Colter, Willard, and Shannon had taken an enormous risk, but they had survived! If there were a trading post somewhere near the ocean, or a ship, one of them would find it. Finally the party was on the move once again.

Now the encounter took an unexpected turn. As the Indians remained in Lewis and Clark's camp, apparently answering more questions about what lay farther downriver, suddenly some rustling in the bushes from the hillside above camp drew everyone's attention, and moments later John Colter tumbled and slid down the hill and back into camp. His rifle had broken, so he had parted ways with Willard and Shannon and returned to camp on foot.[10]

> *at this time one of the men Colter returnd by land* Moulton, *Journals* 6:47 (Clark)

It might have appeared at first that Colter was bringing back a report of what he had seen downriver, when in fact, he happened to be in hot pursuit of the same Indians that were conversing with the captains. Colter quickly explained that he had been robbed and was certain that these were the Indians who had taken his belongings. Of course, everyone could see that the three half-naked Indian men were not hiding anything beneath their small cedar-bark capes, so Captain Clark turned his attention to the two women in the canoe. Perhaps they were hiding Colter's belongings. In order to investigate, he ordered the women to come ashore.

The women were defiant. Instead of bringing their canoe to shore, they remained in deep water where they knew Clark couldn't reach them.

> *I called to the Squars to land and give back the gigg, which they would not doe* Moulton, *Journals* 6:47 (Clark)

This was not a joke, and none of Lewis and Clark's men were in the mood for games. Upon seeing these women ignore their captain's orders, one of the men grabbed his rifle, then ran aggressively to the water's edge and took aim.

untill a man run with a gun, as if he intended to Shute them when they landed, and Colter got his gig & basket Moulton, *Journals* 6:47 (Clark)

Exactly as he suspected, Colter's spear and basket were in the canoe. These items were quickly recovered, and Clark rudely dismissed the thieves. His patience was wearing thin.

I then ordered those fellow off, and they verry readily Cleared out they are of the War-ci-a-cum Moulton, *Journals* 6:47 (Clark)

With this unpleasant encounter behind them, Lewis and Clark now turned their attention to Colter. They had many questions for him. Where had he been? What had he seen? What lay further ahead?

Colter informed us that "it was but a Short distance from where we lay around the point to a butifull Sand beech Moulton, *Journals* 6:47 (Clark)

Colter's report assured them that a canoe could make it around the point. So Lewis now abandoned his plan to go on foot and instead ordered their best canoe to be made ready. Realizing that it would require a lot of manpower to paddle against the surge of the ocean, he selected nine men to accompany him. If they made it, he would split up the party and send

"Those Scoundals"

Lewis and Clark encountered many different Indian tribes as they descended the Columbia River. The various nations reacted differently to the party: some treated them with great kindness;[1] others were frightened and hid.[2] Most of these encounters were pleasant; however, Lewis and Clark became annoyed when they arrived among the Indians of the Lower Columbia because of their tendency to steal.

Why did these Indians steal? Some people explain that they were merely exacting a toll for allowing Lewis and Clark to pass down their river.

we Smoked with them and treated them with every attention & friendship. dureing the time we were at dinner those fellows Stold my pipe Tomahawk which They were Smoking with[3] (Clark)

Others speculate that the Indians viewed everything as common property, without any particular ownership. If they saw something and no one was using it, they took it.

while Serching for the Tomahawk one of those Scoundals Stole a Cappoe of one of our interpreters, which was found Stufed under the root of a treer, near the place they Sat[4] (Clark)

Still others view this as the Indians' way of exacting revenge for having been robbed, cheated, and abused by the fur traders who preceded Lewis and Clark into this area.

Three Indians followed us they Could Speake a little English, they were detected in Stealing a knife & returned late to their village[5] (Clark)

The list of possibilities could go on, and we'll never know exactly what motivated this group of Indians. However, we do know that since Lewis and Clark were already short on supplies, this theft was irritating and forced them to constantly be on their guard whenever Indians were near their camp.

where I had formed a camp on ellegable Situation for the protection of our Stores from Thieft, which we were more fearfull of, than their arrows[6] (Clark)

the canoe back with five men. By midafternoon they were ready to go.

> *at 3 oClock he Set out with 4 men Drewyer Jos. & Reu. Fields & R. Frasure, in one of our large canoes and 5 men to Set them around the point on the Sand beech* Moulton, *Journals* 6:47 (Clark)

Clark and his men waited and watched. One hour passed, then another. Finally, just before nightfall, the canoe returned.

It had been a harrowing trip back upriver and around the point. The canoe was repeatedly hit with waves that splashed and rolled over its side.

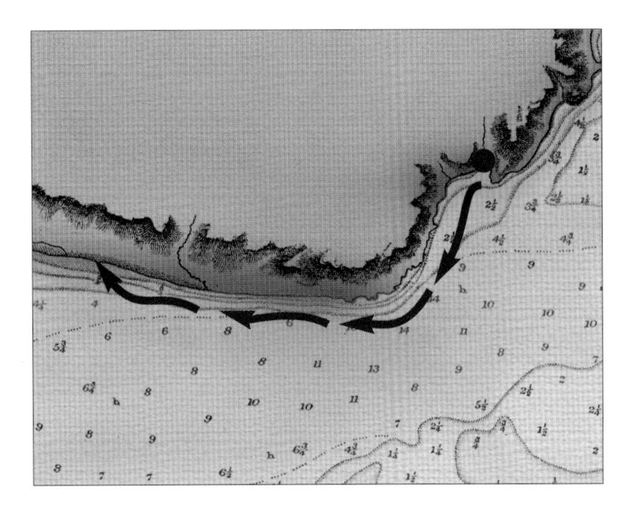

Lewis and nine men take a canoe around Point Distress.

Water swirled around the men's legs as it filled deeper and deeper. The crew knew that at any moment the canoe could swamp and capsize, so they paddled harder and harder to stay in front of the next wave.

> *this canoe returned nearly filled with water at Dark which it receved by the waves dashing into it on its return, haveing landed Capt. Lewis & his party Safe on the Sand beech* Moulton, *Journals* 6:47 (Clark)

Seven men were now around the point. Even though most of the party remained pinned down in the same little cove, they all must have felt elated to know that the party was moving forward once again.

Clark knew he had to get his men out of there as soon as possible. They were suffering terribly from the constant cold rain and miserable, sleepless nights; and now, adding to these problems, the important source of food that swam in the creek was nearly wiped out.

> *The rain Continue all day all wet as usial, killed only 2 fish to day for the whole Party*
>
> Moulton, *Journals* 6:46 (Clark)

A new concern now emerged that was even more troubling than hunger. The men's clothing was falling apart. Leather will almost dissolve if it remains wet, and these clothes hadn't seen a moment of dryness in over a week. Sleeves fell off the men's shirts, and pant legs ripped from the cuff to the waist as if made of wet cardboard. Chunks of rotten leather lay strewn on the ground.

> *The rain &c. which has continued without a longer intermition than 2 hours at a time for ten days past has distroyd. the robes and rotted nearly one half of*

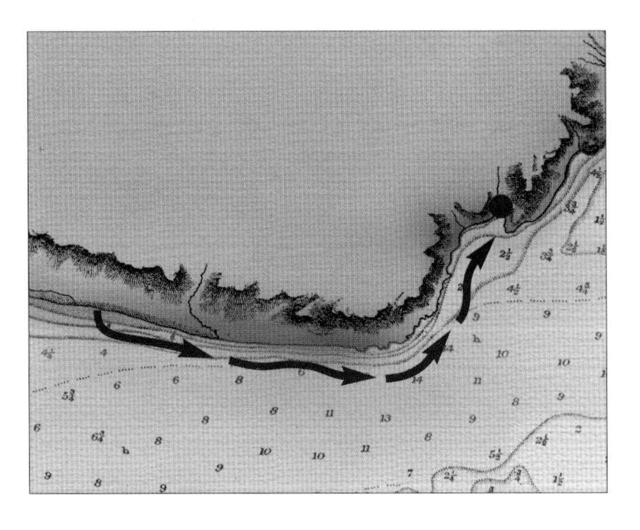

(right) Several hours later, five of the men return to Dismal Nitch. Captain Lewis and four others remain downriver.

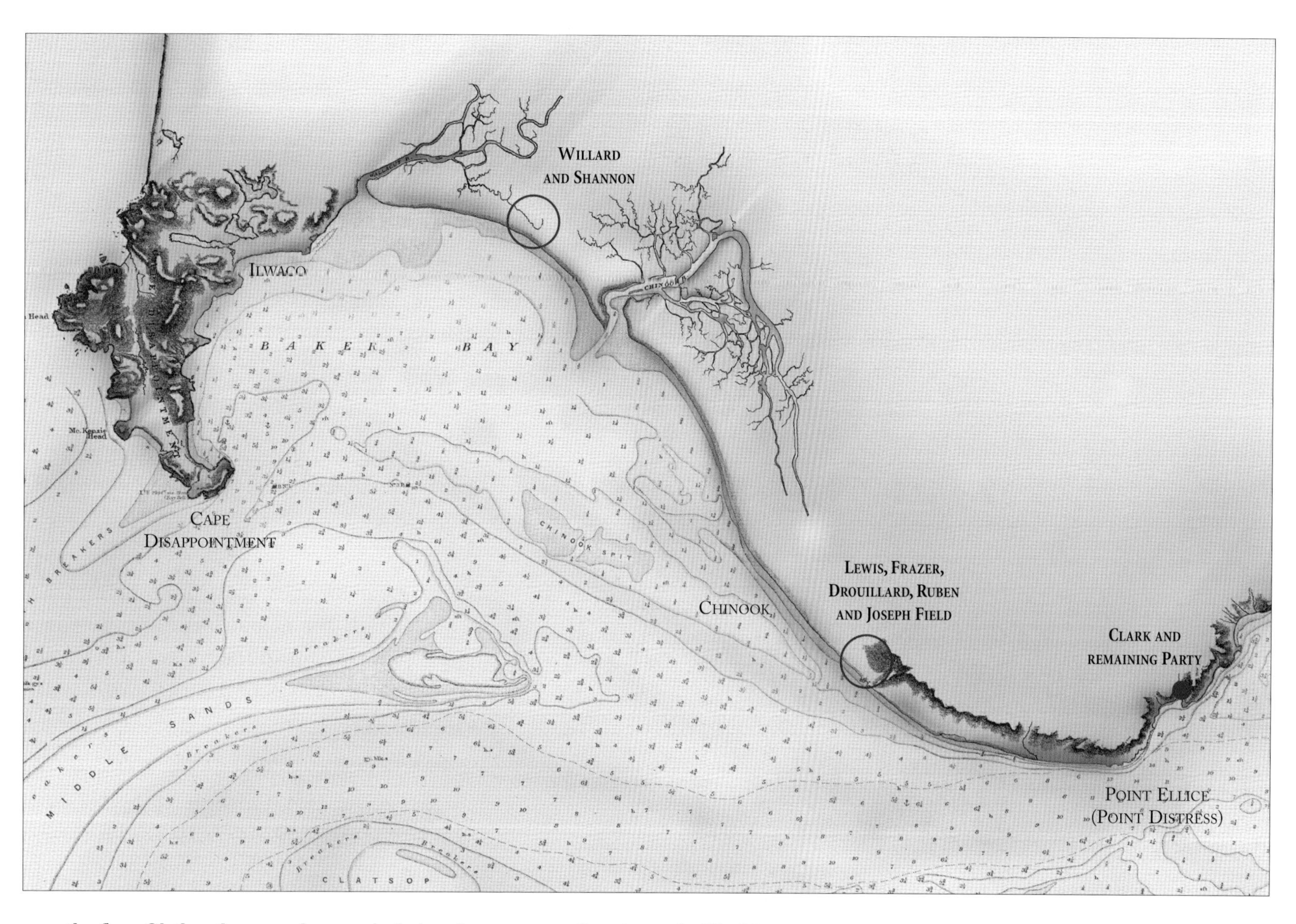

> *the fiew Clothes the party has, perticularley the leather Clothes* Moulton, *Journals* 6:47 (Clark)

Clark was concerned about this grim situation. Winter had arrived. This was no time to have human flesh exposed to the elements. If the temperature were to suddenly drop below freezing, many of his men would be in agony.

> *if we have Cold weather before we Can kill Dress Skins for Clothing the bulk of the party will Suffer verry much* Moulton, *Journals* 6:47 (Clark)

Lewis and Clark's party were now separated into three different groups, with miles in between them. All in all, their attempt to arrive at the ocean had caused them more grief and disappointment than any other part of their journey. They had never spent so many days advancing so few miles.

At nightfall the Lewis and Clark party is now split up into three groups, separated from each other by several miles.

CHAPTER THREE

Arrival at the Ocean

in full view of the Ocian

Friday, November 15th

The jet stream pushes north, effectively holding low pressure systems out of the region. Cold air and partial clearing with sun-breaks dominate coastal weather.

Daytime High Tide:	*8:24 am*	*8.5*
Daytime Low Tide:	*3:06 pm*	*1.7*
Sunrise:	*7:15 am*	
Sunset:	*4:48 pm*	

A SINGLE DROPLET OF WATER IS PRACTICALLY silent; however, when billions of raindrops fall hour after hour in the middle of the woods, each one dripping down though the branches and splashing onto ferns and fallen leaves, the combined effect sounds like a thousand toy drums. Lewis and Clark's men had lived with this constant racket for the past ten days, but on this night it finally came to a stop. The resulting eerie silence would be enough to startle anyone awake. Without that steady drumming noise, the unexpected stillness of the night causes one's ears to go on high alert, wondering at the meaning of each distinct and distant sound.

Calm weather had arrived over the coast for the first time in the past ten days. This was exactly what Clark had been hoping for.

> *Rained all the last night, this morning it became Calm and fair, I preposed Setting out, and ordered the Canoes Repared and loaded* Moulton, *Journals* 6:49 (Clark)

Once again, however, the mood of the Columbia changed and stopped Clark in his tracks.

> *before we could load our canoes the wind Sudenly Sprung up . . . and blew with Such violence, that we could not proceed in Safty with the loading*
>
> Moulton, *Journals* 6:9 (Clark)

Clark must have been dumbfounded by this sudden change. He tried the point alone in a canoe, exactly as Lewis had done, but the waves slapped him sideways and forced him back to camp.

> *I went to the point in an empty canoe and found it would be dangerous to proceed even in an empty Canoe* Moulton, *Journals* 6:48 (Clark)

It was almost as if the Columbia River were trying to purposely lure the men into an ambush. One moment it was calm, then suddenly it turned rough and violent. Clark grew frustrated and perhaps a little testy. They had

(below left) A sudden wind keeps Clark from proceeding downriver.

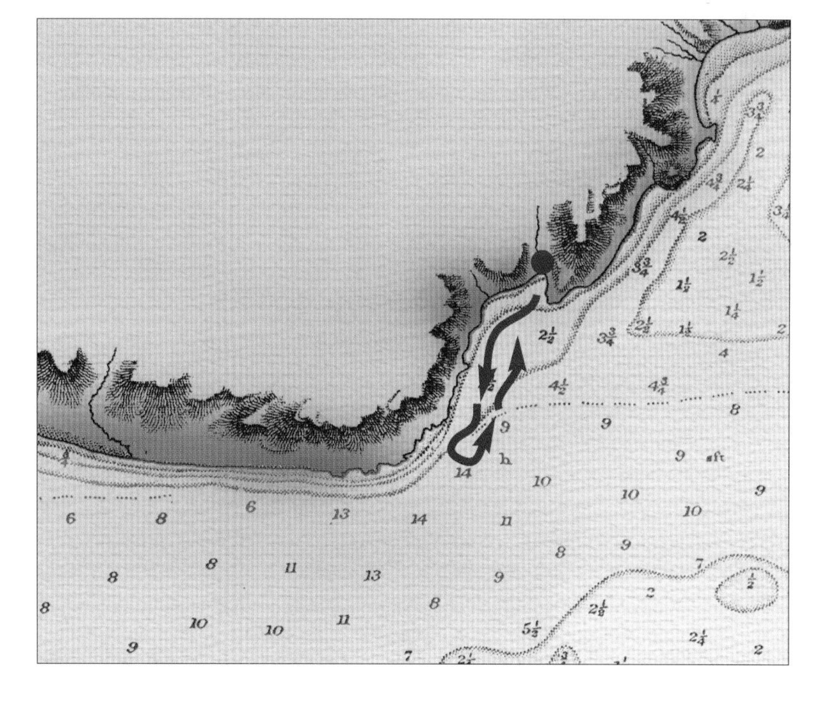

(below) Clark makes an attempt around Point Distress in an empty canoe, but the high waves turn him around.

The single point of rock that stopped the Lewis and Clark party for five consecutive days had been given several names by Clark. He refers to it as "blustering point" and he calls it "Point Distress." Once he referred to it as "Stormey point."

In later years, during the British occupation of Astoria, the point of land was named "Point Ellice."

This historic landmark exists today in virtual obscurity. Most people know it simply as the northern end of the Astoria Bridge.

been in bad spots before, but this was absolutely the worst.

> *the most disagreeable time I have experienced . . . where I can neither get out to hunt, return to a better Situation, or proceed on* Moulton, *Journals* 6:48 (Clark)

While all of this was occurring, miles and miles away Willard and Shannon were having a terrible morning as well. The night before, they had camped with friendly Indians and had taken every precaution to guard themselves against theft. They had even placed their rifles beneath their heads like pillows. Yet somehow, in the middle of the night, their rifles mysteriously disappeared. When they awoke and discovered that the rifles were missing, they immediately confronted the Indians and warned them that if the rifles were not given back, a large group of their companions would soon arrive to punish them. The Indians, however, were unmoved by the idle threats from these harmless, now unarmed, men. Dejected and defenseless, Willard and Shannon retreated, walking back upriver along the shore.

The two men assumed that they would have to walk eight miles back to Point Distress, during which time the Indians would make their escape and they would never see their rifles again. Fortunately, Lewis's party was nearby and quickly coming their way. After just a short walk, the two groups ran into each other in what must have been a joyful reunion.

Upon hearing their story, Lewis and his party charged into the Indian camp, caught the culprits, and persuaded them to give back the rifles. Whether this was done with diplomacy or with a threat of violence is unknown, but the rifles were returned.

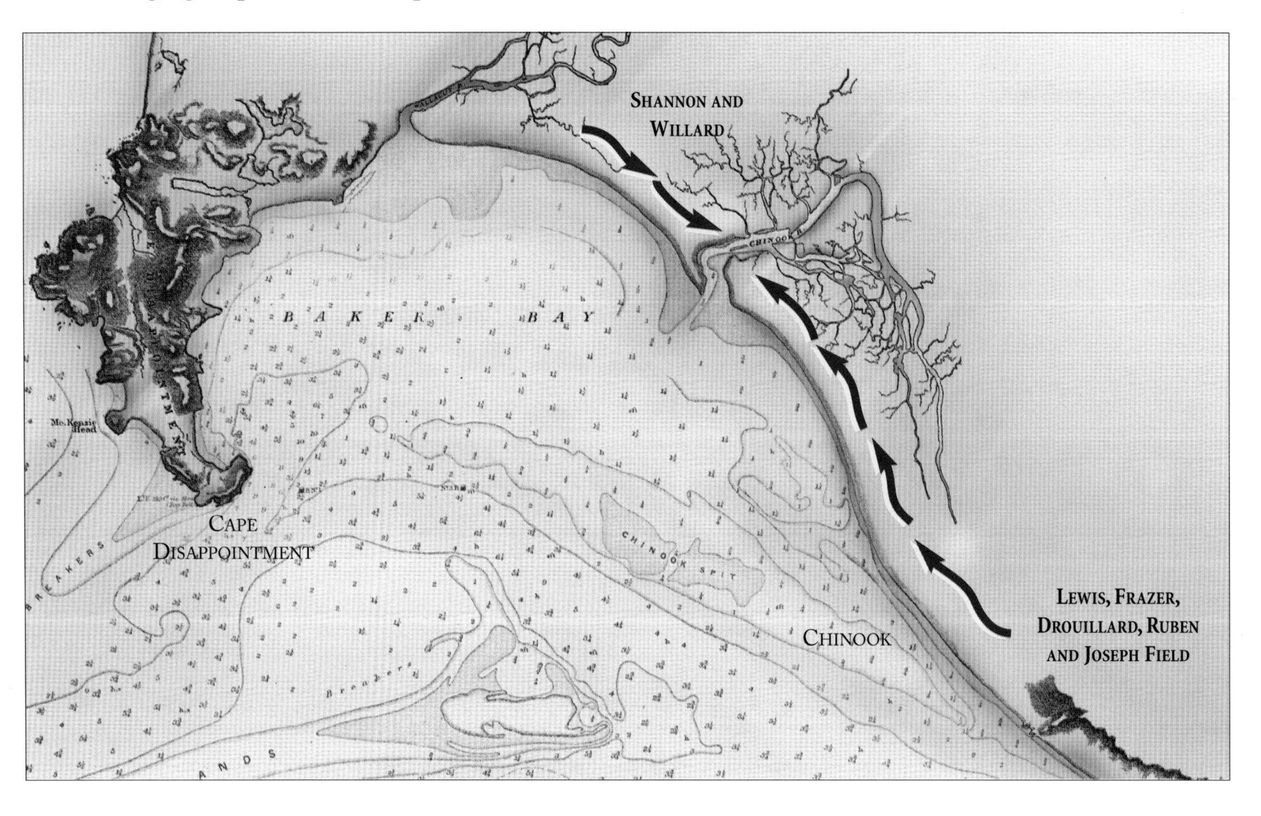

Willard and Shannon's search for fur traders is interrupted when they awaken to discover their rifles have been stolen. They decide to join up with the main body of the party, perhaps to get help. They do not realize that Lewis is heading in their direction.

This was a dicey moment for Lewis. He didn't know whether this robbery was an isolated bit of mischief, or whether the Indians were planning more assaults. He wanted to continue his search for sailing ships around Cape Disappointment, but at the same time he wanted to warn Clark about the thieving Indians so that he wouldn't let down his guard. Lewis decided to split up the two men. Shannon was sent back to

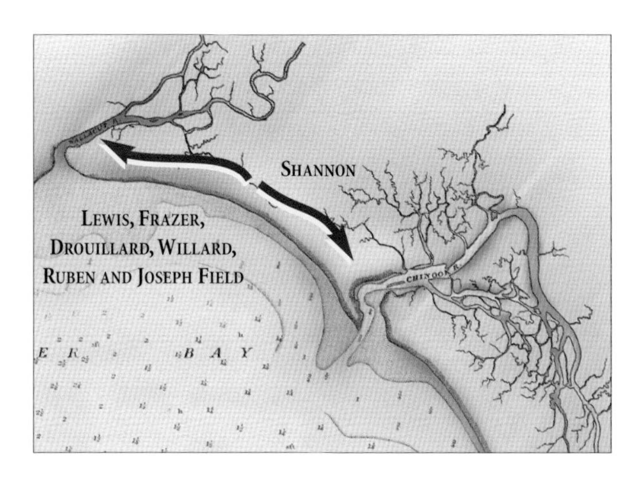

warn Clark, while Willard stayed with him, perhaps to increase his party from five to six men, in case another altercation occurred.

Already, within the first couple of hours of daylight, there had been more excitement and drama than these men had seen in weeks. In addition to this, there now appeared a significant change in the weather. The clouds lifted and sunshine attempted to burn through. Clark and his men remained stuck, but at least now they could dry their clothes. Clark also kept the party busy cleaning their rifles and inspecting their supplies.

> *The Sun Shown untill 1 oClock p.m. which gave an oppertunity for us to dry Some of our bedding, & examine our baggage, the greater Part of which I found wet*
>
> Moulton, *Journals* 6:48 (Clark)

While all of this was occurring, the great Columbia River was ebbing. Hour after hour, billions and billions of gallons of fresh water poured out into the ocean, lowering the level of the river inch by inch. Finally, after six hours, the current began to slow. It was *low water slack*, and the Columbia flattened into a smooth, blue sheet. Clark saw this opportunity to get around the point and seized the moment. He ordered his men into action.

> *About 3 oClock the wind luled, and the river became calm, I had the canoes loaded in great haste and Set Out, from this dismal nitich*
>
> Moulton, *Journals* 6:49 (Clark)

(left) Lewis retrieves the rifles and sends Shannon back upriver to warn Clark about the robbery. He keeps Willard with his party as they continue their search for fur traders around Cape Disappointment.

They paddled along the black rock shore. The waves were gone; their canoes glided quickly through the calm water. Within minutes, they were around the point. Clark sums up this passing with just one line.

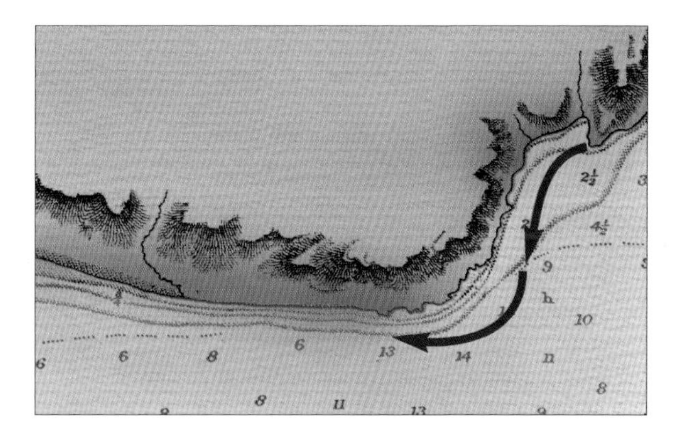

The low tide and calm wind give Clark a chance to get around Point Distress. His entire party, along with all their baggage, set out in the four large canoes and quickly slip around the point without mishap.

> *proceeded on passed the blustering point*
>
> Moulton, *Journals* 6:49 (Clark)

Just that quickly the mouth of the Columbia came into view. And there, immediately beyond, was the great Pacific Ocean. This was the party's first close-up view of these waters.

Exactly as Colter had described, they encountered a long sandy beach.

I found a butifull Sand beech Moulton, *Journals* 6:49 (Clark)

Directly ahead of them was an enormous village of wooden houses. It was the largest village they had seen along the entire lower Columbia, but, oddly enough, every house was empty.

below the mouth of this Stream is a village of 36 houses uninhabited Moulton, *Journals* 6:50 (Clark)

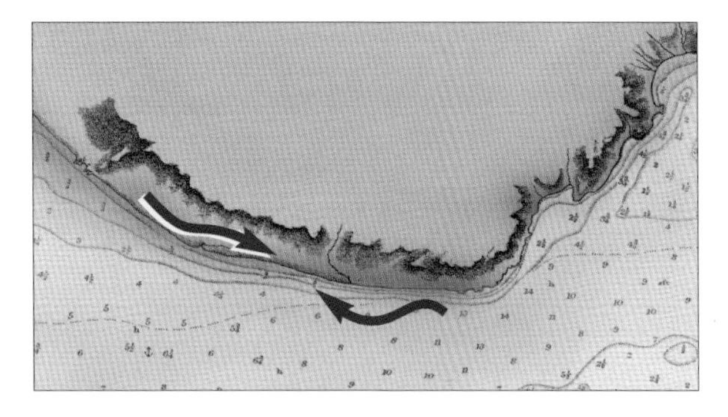

Clark meets Shannon who tells him about the incident with the thieving Indians.

Even more surprising was the sudden appearance of George Shannon walking along the shore with five Indians. Where was Willard? What was going on? Clark pulled in close to the beach and soon learned what had happened.

Shannon informed me that he met Capn. Lewis . . . who had Sent him back to meet me, he also told me the Indians were thievish, as the night before they had Stolen both his and Willards rifles from under their heads Moulton, *Journals* 6:50 (Clark)

Clark was outraged by this news. Shannon continued with his story.

they Set out on their return and had not proceeded far up the beech before they met Capt Lewis

Moulton, *Journals* 6:50 (Clark)

Capt. Lewis & party arrived at the Camp of those Indians at So Timely a period that the Inds. were allarmed & delivered up the guns

Moulton, *Journals* 6:48 (Clark)

Clark's patience had been worn razor-thin by the weather, hunger, and misery of these past few days. He had not slept an entire night in over a week. The theft of his men's rifles was more than he could endure. In a rage, Clark confronted the five Chinook Indians accompanying Shannon and made it clear that he would put to death the next Indian who stole anything from his men.

I told those Indians who accompanied Shannon that they Should not Come near us, and if any one of their nation Stold anything from us, I would have him Shot Moulton, *Journals* 6:50 (Clark)

It is unlikely that any of these particular Indians were aware of the theft. They were probably just curious locals who saw these white men wandering back and forth and wanted to find out where they came from, perhaps in hopes of negotiating some trade. Now, suddenly, they found themselves having their lives threatened. These Chinooks must have been completely baffled.

The daylight was quickly slipping away. The tide had now switched and was flooding in from the ocean. Clark led their canoes on downriver, lumbering through the heavy surf until they arrived at a low, sandy point. From here, Clark had a commanding view of the ocean, which was framed on one side by Cape Disappointment and on the other by Point Adams. What's more, from this exact spot he could turn eastward and see miles and miles upriver. In other words, this place gave an extensive view of the entire lower river and appeared to be an ideal location for a camp.

I landed and formed a camp on the highest Spot I could find Moulton, *Journals* 6:50 (Clark)

President Jefferson had sent them across the continent to find the most direct route to the ocean, and there the ocean was, right before them. Clark could now describe, both with maps and words, every mile between the mouth of the Missouri and the ocean. He could describe every curve, every rapid, every point of land, and every waterfall. There was no need to travel another mile. They had arrived.[1]

this I could plainly See would be the extent of our journey by water . . . in full view of the Ocian

Moulton, *Journals* 6:50 (Clark)

The men quickly unloaded their canoes and set up camp. Lumber from the Indian village was reassembled into crude shelters where the men could sleep. This would be their home by the ocean; Clark called it Station Camp.

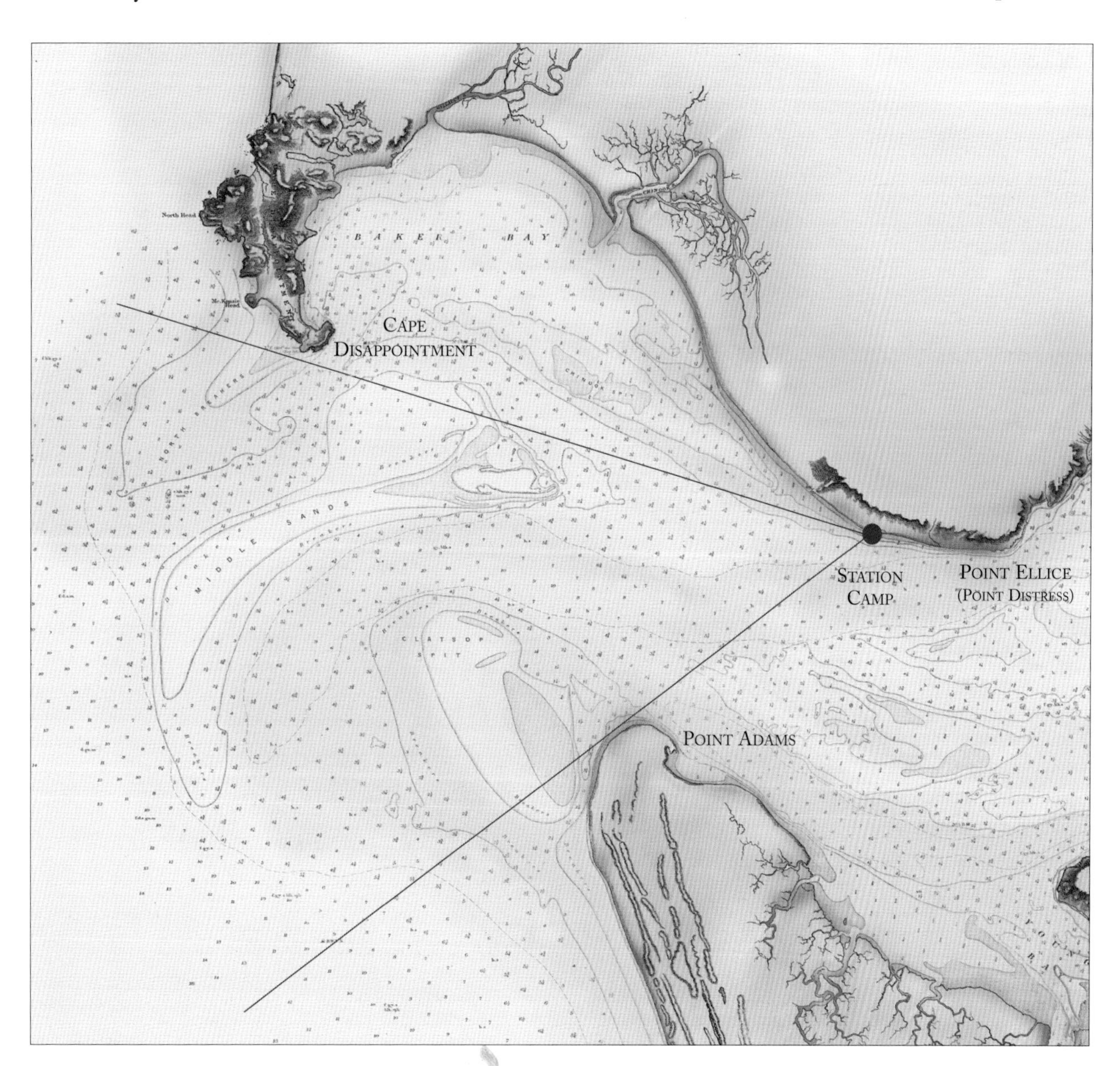

"In full view of the Ocian," Clark writes in his journal. The party unloads their canoes and constructs shelters at the place they will call Station Camp.

From Station Camp the men look directly into the Pacific Ocean.

our men all Comfortable in their Camps which they have made of boards from the old Village above

Moulton, *Journals* 6:50 (Clark)

Several more Indians arrived, offering roots for sale, but Clark did not show them any kindness. To make sure that there was no misunderstanding about why he was so angry, he addressed them once again to explain the cause.

The Indians who accompanied Shannon from the village below, . . . Call themselves Chin nooks, I told those people that they had attempted to Steal 2 guns &c. that if any one of their nation stole any thing that the Sentinl. whome they Saw near our baggage with his gun would most certainly Shute them

Moulton, *Journals* 6:50 (Clark)

These Indians didn't want trouble; they were merchants who were only interested in trading, and turning a tidy profit. So in order to appease this white man and get on with the trading, they acknowledged Clark's threat and agreed to punish any wrongdoers.

they all promised not to tuch a thing, and if any of their womin or bad boys took any thing to return it imediately and Chastise them for it

Moulton, *Journals* 6:50 (Clark)

The relationship between Clark and the Chinooks had been poisoned. There was nothing they could do or say that would ever mend this rift.

> *I treated those people with great distance*
>
> Moulton, *Journals* 6:50 (Clark)

Lewis and Clark and their men were finally at their destination. Reaching this goal had taken them eighteen months. One would assume there would have been a celebration; however, this arrival had in fact been anticlimactic. There was no backslapping or joyfulness. The men had just gone through one of the worst ordeals of their lives and were bony, exhausted, miserable wrecks, who just felt thankful to be alive. In appearance, Clark and his ragged men probably resembled prisoners of war.

Evening was upon them; the sky quickly turned dark. They dried their blankets and crawled inside their shelters for what would be their first comfortable sleep in more than a week. The ocean roared like thunder, but every one of Clark's men must have slept right through until dawn.

Saturday, November 16th

The weather continues calm and fair. Sunbreaks.

Daytime High Tide:	*9:33 am*	*8.5*
Daytime Low Tide:	*4:08pm*	*1.0*
Sunrise:	*7:16 am*	
Sunset:	*4:47 pm*	

THE WEATHER WAS NOW THE EXACT OPPOSITE of what it had been during the past several days. Instead of enduring a night of wet blankets with water wicking up from below and torrential downpours from above, Clark and his men enjoyed the first dry night of uninterrupted sleep in more than a week.

> *Cool the latter part of the last night this morning Clear and butifull* Moulton, *Journals* 6:53 (Clark)

Clark's first thought was to get their camp in order. He directed the men to have all of their supplies unpacked and inspected, and, as one might expect after so many days of constant rain, everything was found to be soaking wet. The men sorted through the packets of medicine, navigational instruments, tools, blankets, books, and scientific specimens. It must have been a mess. With all their blankets and clothes strewn across every driftwood log, the camp probably resembled a shipwreck that had washed ashore.

> *I had all our articles of every discription examined and put out to Dry* Moulton, *Journals* 6:53 (Clark)

The men were still hungry. They had not had a good hot meal in over a week, so Clark ordered every available hunter out into the woods.

> *I Sent out Several hunters and fowlers in pursute Elk, Deer, or fowls of any kind* Moulton, *Journals* 6:53 (Clark)

Clark then turned his attention to the geography of the mouth of the Columbia. He examined the shorelines of both sides of the river and jotted down a brief description of the hills and forests that he could see.

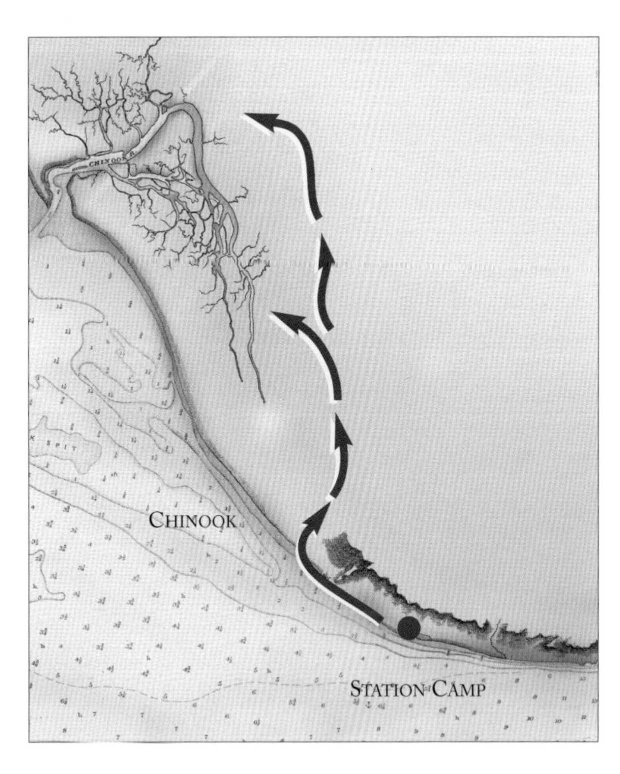

Clark sends out every available hunter to shoot as much game as possible for his famished men.

> *The Countrey . . . above Haley Bay is high broken and thickley timbered . . . from Point Adams the Contrey appears low for 15 or 20 miles back to the mountains* Moulton, *Journals* 6:54 (Clark)

The landscape that surrounded this campsite could not hold one's attention very long, with that wild, magnificent body of water relentlessly pounding the shore. There it was, the great Pacific Ocean. Breakers curled and slapped along the beach in front of their camp, tossing a salty spray into the air. The hypnotic motion of the waves must have been a source of constant fascination for Clark's men.

> *We could see the waves, like small mountains, rolling out in the ocean* Moulton, *Journals* 10:171 (Gass)

They were now face to face with their destination, and the men knew it. Joseph Whitehouse's journal plainly reveals that, in the opinion of the

"We could see the waves, like small mountains, rolling out in the ocean"

Patrick Gass

Clark often refers to York as his "servant." Legally, Clark owned him, which makes him a slave. From this point forward in the text, York will be referred to as Clark's "servant."

members of the party, they were at the western end of their journey. He wrote:

> *We are now of opinion that we cannot go any further with our Canoes, & think that we are at an end of our Voyage to the Pacific Ocean*
>
> Moulton, *Journals* 11:394 (Whitehouse)

Sergeant Gass goes a step further. His journal shows that the men were fully aware of President Jefferson's instructions and that they knew they had fulfilled those orders to the letter (*see Appendix Four: "Jefferson's Letter to Lewis, 1803," p. 196*). Every curve and every mile of waterway between the ocean and the Mississippi had been explored. The mission was a success. Their work was over.

> *We are now at the end of our voyage, which has been completely accomplished according to the intention of the expedition, the object of which was to discover a passage by the way of the Missouri and Columbia rivers to the Pacific ocean*
>
> Moulton, *Journals* 10:171 (Gass)

They had achieved their goal. From here they would turn around and begin their homeward journey.

The hunters returned in the evening with bad news. The big game animals, such as elk or bear, had completely eluded them. They were, however, successful at shooting a couple of small deer and a dozen geese and ducks. While this would not be enough meat for a feast, it was sufficient for one good, hot, fresh meal – their first in over a week. It is interesting to note that Clark sent his servant York out hunting and that he shot the majority of the game.

> *our hunters and fowlers killd 2 Deer 1 Crane & 2 ducks, my Servt. York killed 2 Geese & 8 white, black and Speckle Brants* Moulton, *Journals* 6:53 (Clark)

Captain Lewis and his party, meanwhile, continued their search for the rumored fur traders' camp. They searched all around Cape Disappointment looking for any trace of them, but because Lewis did not keep a journal during this excursion, their precise movements cannot be confirmed.

The weather remained calm throughout the night. Lewis and Clark's two camps were separated by a dozen miles, but they shared the same clear, pleasant sky.

Where are the Chinooks?

Most Indians did not live along the shore of the Columbia River all year, and this was particularly true of the Chinooks. Each Chinook family had two or more houses, and they moved from one to another depending on the season.

Typically, the Chinooks would leave their villages along the shore of the Columbia after the salmon migration was finished, but before winter storms arrived. This movement away from the river occurred around October.

During the dark, stormy months of November through January, the Chinooks resided in the winter villages, well protected from the high winds. Then, beginning in January, some returned to their homes along the Columbia in order to prepare for the February smelt fishing. Others followed, and by April the entire population was settled into their Columbia River villages, ready for the salmon that arrived in May.

The short migration route was covered by canoe. The Chinook could ascend one river, make a short portage, and descend another stream without any complications.[1] The move was accomplished in one day.

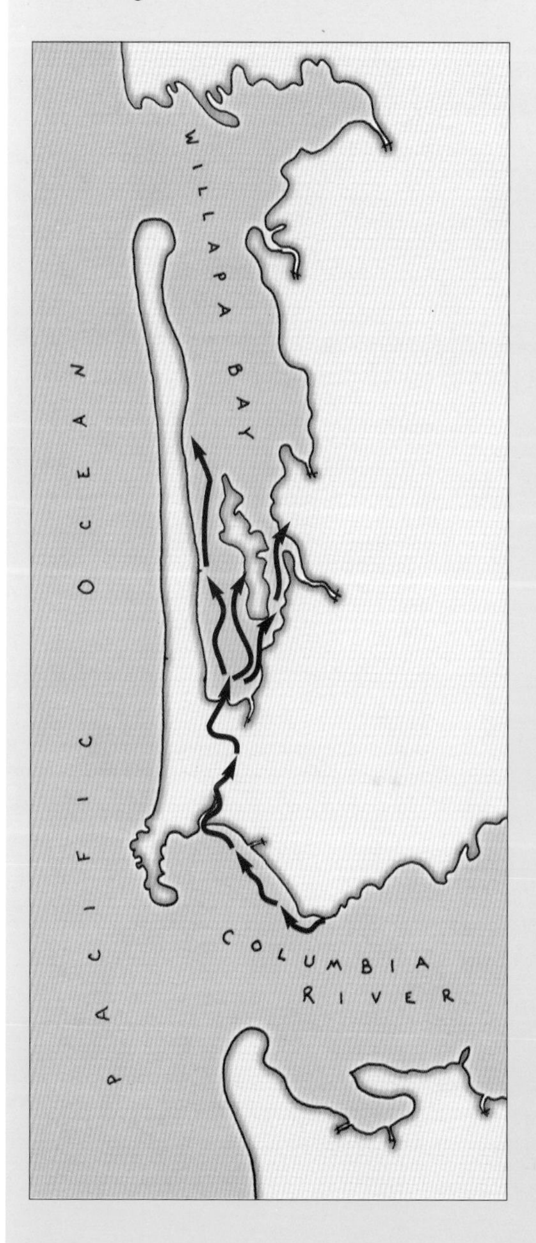

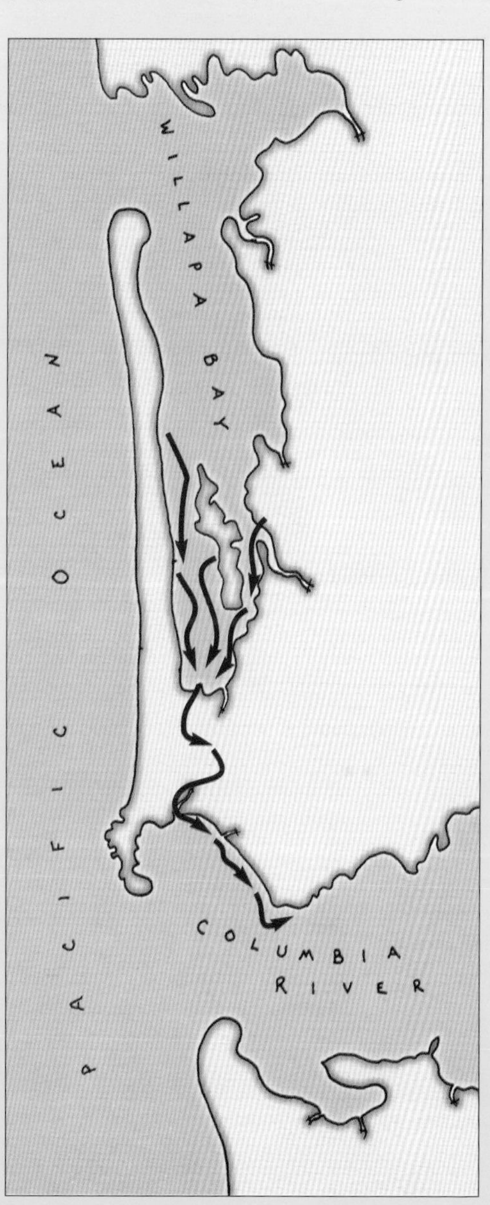

(*left*) Gabriel Franchere, of the Astor Party, described this migration away from the river in early October of 1811:

> *Meanwhile, the season being come when the Indians quit the seashore and the banks of the Columbia to retire into the woods and establish their winter quarters along the small streams and rivers, we began to find ourselves short of provisions.*[2]
>
> *Experience having taught us that from the beginning of October to the end of January provisions were brought in by the natives in very small quantity.*[3]

(*right*) In 1814 Alexander Henry, of the Northwest Company, observed the return of the Chinooks to the Columbia in late January.

> *The great smoke which rises from the three Chinook villages denotes the return of the people, as usual at this period; they will increase in numbers daily, as smelt-fishing is approaching fast; sturgeon-fishing follows, and then salmon-fishing, as spring draws near.*[4]
>
> *The three villages on Chinook point seem to have increased much in population, if we may judge from the smoke which rises from them.*[5]

Sunday, November 17th

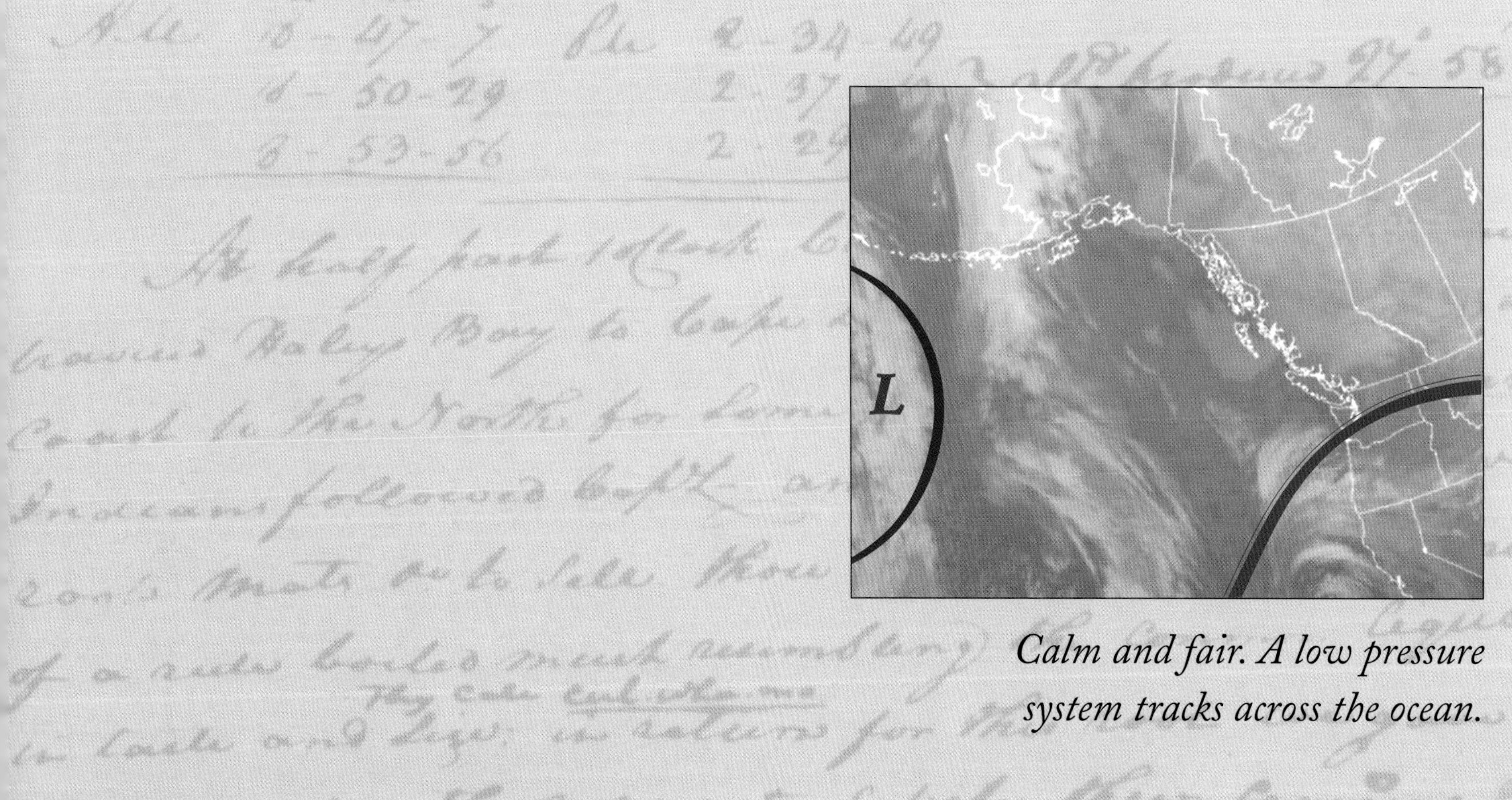

Calm and fair. A low pressure system tracks across the ocean.

Daytime High Tide:	*10:19 am*	*8.9*
Daytime Low Tide:	*5:03 pm*	*0.3*
Sunrise:	*7:17 am*	
Sunset:	*4:46 pm*	

Lewis and Clark's men might have been perplexed by this changeable weather. A couple of days earlier they could hardly stand up because of the high winds and driving rain; now, in direct contrast, the weather was calm and pleasant. They probably wondered if all that violent weather hadn't been some sort of rare hurricane blowing through the region. Not a single droplet of rain had fallen in the past forty-eight hours, and the horizon was clear.

> *A fair cool morning wind from the East*
>
> Moulton, *Journals* 6:61 (Clark)

The main party was resting and recuperating, but lack of food was still a problem, so Clark sent his best hunters into the woods to bring in as much meat as possible.

> *I Sent out 6 men to kill deer & fowls this morning*
>
> Moulton, *Journals* 6:60 (Clark)

Clark continued with his own work of measuring and describing the land surrounding the mouth of this great Columbia River. Nothing escaped his attention. He even measured, quite accurately, the rise and fall of the tidewater.

> *The tide rises at this place 8 feet 6 inches and comes in with great waves brakeing on the Sand beech on which we lay* Moulton, *Journals* 6:61 (Clark)

Mapping this vast river's mouth was an enormous challenge for Clark. The distant shore was miles away, and before he could begin drawing, he would first have to pinpoint the location of certain significant points. This required triangulations with his compass, so he walked back upriver to Point Distress, drove a stake into the sand, then meticulously measured the entire length of the shoreline. The distance from Point Distress to the Chinook village was 3,861 feet, and from this village to their camp was another 2,673 feet. He took a second set of readings from camp, and farther downriver a third. Clark approached this task with the seriousness of an engineer.[2]

> *I Surveyed a little on the corse and made Some observns* Moulton, *Journals* 6:60 (Clark)

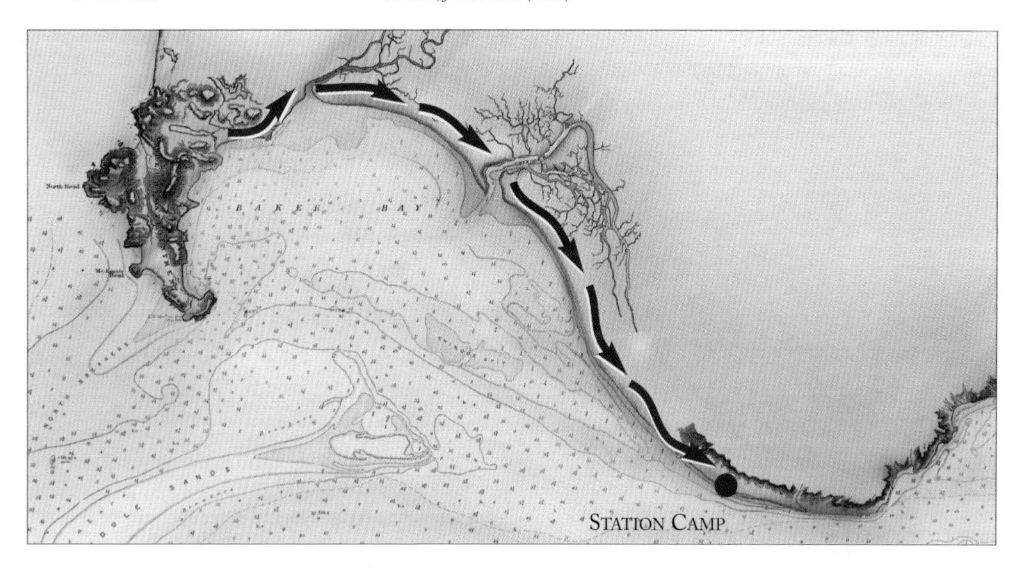

Lewis and the five men accompanying him return from Cape Disappointment, their search for fur traders unsuccessful.

In the early afternoon Captain Lewis and his party returned. He had spent three days searching for fur traders around Cape Disappointment, but had found nothing. Clark summed up this reunion with a single line:

> *at half past 1 oClock Capt. Lewis and his Party returned haveing around passd. Point Disapointment and Some distance on the main Ocian to the N W.* Moulton, *Journals* 6:60 (Clark)

Clark's failure to mention the rumored fur traders is a mystery. Having written about "white men" during the past two months, why didn't he write anything more about Lewis's discouraging report? Fortunately, other members of the party wrote about Lewis's return in their journals, giving us a clearer idea of what Clark would have heard. Whitehouse noted:

> *Captain Lewis, and the Men that was out with him also returned. They informed us, that they had been about 30 Miles down on the Sea Coast, & that they had seen no white people or Vessells. They learnt from the Indians along the Coast that some white people & Vessells had been lately there but that they were all gone.* Moulton, *Journals* 11:395 (Whitehouse)

Now it was clear to everyone that an encounter with a sailing ship was not going to happen. There would be no new blankets, clothes, shoes, or canvas tents. What's more, they had no way to send copies of their journals and maps to President Jefferson, and all the scientific collections of plants, skins, bones, seeds, horns, and feathers would have to be carried back with them.

Also, Lewis and Clark's party was rapidly approaching bankruptcy. The supply of beads, mirrors, fishhooks, and blankets they had used as currency among the Indians was nearly exhausted. They had overspent on the westward leg of their journey, and without any additional supplies to exchange for food, guides, and information, their homeward journey would be difficult.

Naturally, Lewis and Clark wanted to save their remaining trade goods for the long journey home; however, that was not easy to do, especially here among the Chinooks who had dealt with fur traders for more than a dozen years. They had high expectations and were accustomed to lavish gifts, which obviously Lewis and Clark did not have.[3] Several Chinooks followed Lewis back to camp and brought with them some items to sell. Negotiations began with the traditional exchange of gifts, and this soon resulted in a misunderstanding.[4]

> *Several Chinnook Indians followed Capt L— and a Canoe came up with roots mats &c. to Sell. those Chinnooks made us a present of a rute boiled much resembling the common liquorice in taste*
>
> Moulton, *Journals* 6:61 (Clark)

Now, following a well-established tradition, Lewis and Clark were expected to give back

Trade Goods

Trading ships came to the Northwest Coast to buy sea otter pelts. Each ship brought along a wide variety of items for trade and in many ways would have resembled a floating department store.[1]

If Lewis and Clark had met a fur trading ship, they could have outfitted themselves with canvas for tents, new clothes for the men, blankets, and trinkets to trade with the Indians. This replenishment of goods would have made their return trip much easier and more comfortable.[2]

This list of trade items was part of the inventory loaded aboard the ship Atahualpa *before leaving Boston in 1800, bound for the northwest fur trade.*[3]

Broadcloth, 5,796 yards	*Buttons, 600 dozen*	*Chisels, 168*
Greatcoats, 181	*Knives, 400 dozen*	*Saws, 47*
Jackets, 121	*Scissors, 100 dozen*	*Files, 480*
Trousers, 124	*Iron Pots, 301*	*Spoons, 48 dozen*
Blankets, 1,342	*Frying Pans, 240*	*Combs, 30 dozen*
Heavy cotton fabric, 1,980 yards	*Molasses, 623 gallons*	*Looking glass, 771*
Gunpowder, 3,000 units (?)	*Hats, 242*	*Kettles and pans, 570*
Leather flasks, 250	*Iron, Brass, Copper Wire,*	*Canisters, 200 dozen*
Cartouche boxes, 150	*620 units, (pounds, yards, spools)*	*Bottles, 49 dozen*
Muskets, 350	*Axes, 480*	*Beads, $362 worth*
Pistols, 20	*Hammers, 24 Dozen*	*Thread, 30 pounds*

something of equal value. The captains gave what they felt was proper, but the Chinooks were appalled at so small a return. They were expecting much, much more and did not hide their disappointment.

> *in return for this root we gave more than double the value to Satisfy their craveing dispostn. It is a bad practice to receive a present from those Indians as they are never Satisfied for what they reive in return if ten time the value of the articles they gave* Moulton, *Journals* 6:61 (Clark)

The rift between the explorers and the Chinooks now grew wider. After the attempted theft of their rifles, Clark viewed the Indians as rogues and thieves. Now, on top of this, he began to view their trading practices as greedy, fueled by their "craving dispositions." The Indians, on the other hand, probably viewed Lewis and Clark as trigger-happy, overly aggressive, and stingy. The differences between these two groups would never be reconciled.

However, Lewis and Clark could not allow personal feelings and experiences to interfere with their work. Along with the exploration of unknown lands, President Jefferson expected these men to be part-time zoologists, botanists, geologists, and ethnographers. So without further delay, Clark used this opportunity to document what he could learn about the Chinook people and their way of life.[5]

> *this Chin nook Nation is about 400 Souls inhabid the Countrey on the Small rivrs which run into the bay below us on the Ponds to the N W of us* Moulton, *Journals* 6:61 (Clark)

After noting their population and approximate territory, Clark jotted down a few notes about their diet.[6]

> *live principally on fish and roots, . . . and Sometimes kill Elk Deer and fowl.* Moulton, *Journals* 6:61 (Clark)

The weather continued calm and dry. After such a long and grueling journey, Lewis and Clark's men needed this time to rest and recuperate. The hunters returned with more geese, duck, and deer, providing another fresh hot meal. The day passed quietly.

In order to draw a precise map of the river's mouth, Clark needed more measurements, so he decided to extend his survey all the way down to Cape Disappointment. Whitehouse described the decision in this way:

> *Captain Clark concluded to go down with a party tomorrow to the Ocean in order to make his obsersvations of the Coast &c.*
> Moulton, *Journals* 11:395 (Whitehouse)

Clark issued an open invitation to anyone who wanted to accompany him.

> *I directed all the men who wished to See more of the Ocean to Get ready to Set out with me on tomorrow day light* Moulton, *Journals* 6:60 (Clark)

"Men who wished to See more of the Ocean"

Clark listed the members of the party who accompanied him to the ocean. They were as follows:[1]

Sergeant Nathaniel Pryor
Sergeant John Ordway
Peter Weiser
Labiche
Charbonneau
Joseph Field, Reuben Field } Both men had just returned from Cape Disappointment with Lewis.
John Colter, Joseph Shannon } Both men had already been farther downriver.
York } Being Clark's servant, he may have had no voice in the decision.

William Bratton was listed once, but not on the day of the expedition, so it is unclear if he accompanied the group.

Seventeen members of the party and Sacagawea stopped at the final campsite and went no farther west. They were encamped in full view of the ocean and were satisfied with their view.

Exactly how many men Clark thought would accept this offer is unknown; however, he might have been somewhat disappointed by the weak response. Apparently, the majority felt that they had seen enough ocean right in front of their camp and did not need to walk a dozen miles to see more waves.

> *the following men expressed a wish to accompany me . . . all others being well Contented with what part of the Ocean & its curiosities which Could be Seen from the vicinity of our Camp*
>
> Moulton, *Journals* 6:60-61 (Clark)

As darkness fell over the coastline, the men once again settled in for another restful night. This was the first night that Lewis and Clark and all their men were encamped together in full view of the ocean.

Arrival Date

Lewis and Clark first saw the ocean on November 7, but from a distance of twenty-five miles. Lewis got his first close-up view when he rounded Point Distress on November 14; Clark saw the ocean up close on November 15. Now the question, which of these should they choose as the official date of their arrival at the Pacific Ocean?

The first time this arrival date is mentioned is in a letter Lewis wrote as they prepared to leave Fort Clatsop. In this note, he cites November 14 as their arrival date.

> *who were sent out by the government of the U' States in May 1804 to explore the interior of the Continent of North America, did penetrate the same by way of the Missouri and Columbia Rivers, to the discharge of the latter into the Pacific Ocean, where they arrived on the 14th November 1805*[1]

Of course, this excluded Clark and most of the party who remained stuck in Dismal Nitch on that day.

Upon returning to St. Louis, both men worked on a draft of a letter that was reprinted in many newspapers and widely circulated. In this letter November 14 was changed to November 17.

> *On the 17th of November we reached the Ocian where various considerations induced us to spend the winter*[2]

Why they agreed upon November 17 as the official date of their arrival is not clear. Perhaps it was to mark the first day that both captains, along with their entire party, camped together in full view of the Pacific Ocean.

Indian Population

"400 Souls"

President Jefferson asked Lewis and Clark to record the populations of the Indian tribes they encountered.[1] The exact population of the Chinooks in 1805 is unknown; however, Clark estimated their population at a rather modest number:

> *This Chin nook Nation is about 400 Souls inhabid the Countrey on the Small rivrs which run into the bay below us and on the Ponds to the NW of us*[2]

Since these people were in their winter villages at the time, Clark saw only a few dozen tribal members, which undoubtedly accounts for his low population figure. If he had arrived at this campsite two or three months earlier, during the height of the salmon fishing season, it is likely that he would have met between 1,000 and 3,000 individuals, and his census would have been much more accurate.

One possible indication of the size of the Chinook population was revealed in 1813. Chief Comecomly of the Chinooks offered 800 warriors to fight with the Americans against the "King George Men." Perhaps Comecomly was exaggerating, or maybe he was counting alliances with other tribes; but if he were describing only adult male Chinooks, it is possible that their entire population was well over 4,000 individuals.[3]

Speculations on a Map

When Lewis and Clark's journals were first published in 1903, the editor of the journals titled this beautiful map, *Mouth of the Columbia River, sketch map by Clark.* It has been known by this name ever since.[1] This is worth mentioning for two reasons: first, Clark did not give this map that title; and second, this is not a map of the mouth of the Columbia.

Clark was an expert cartographer; his drawings were done with extreme care. If it had been his intention to draw a map of the mouth of the Columbia River, are we to believe he would have selected such a bizarre composition? Why didn't Clark design this map to include both shorelines? Why didn't he draw it with a correct north and south orientation? Who could look at this map and formulate a correct impression of the shape of this river's mouth?

It is obvious that Clark had a different intention when he designed this map. That leads us to this question: what exactly was he trying to show us?

What Clark has done here is to create a sort of *treasure map*. His intention was to pinpoint one specific campsite along the shore. In order to find Station Camp, even today, a person needs only a compass and a copy of this map. By reading the angles Clark precisely recorded and finding where the three lines converge, this camp can be located. Of all the maps Clark drew during his expedition, this is the only one that pinpoints a campsite.

Why did Clark draw a map that directs attention to this one specific site? Clearly he was aware that this successful crossing of the continent had been a great accomplishment. They were the first; this was a historic location.

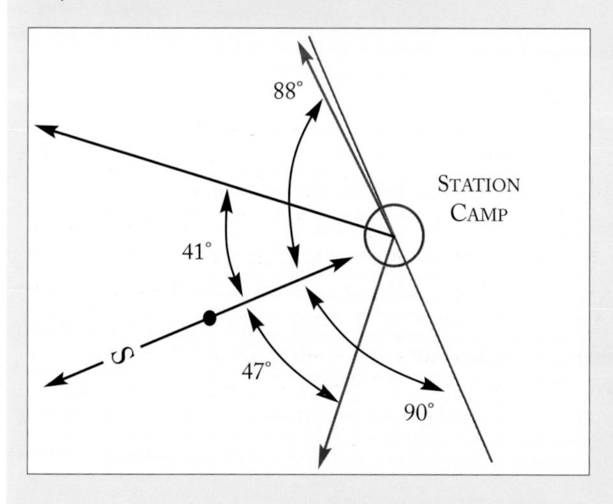

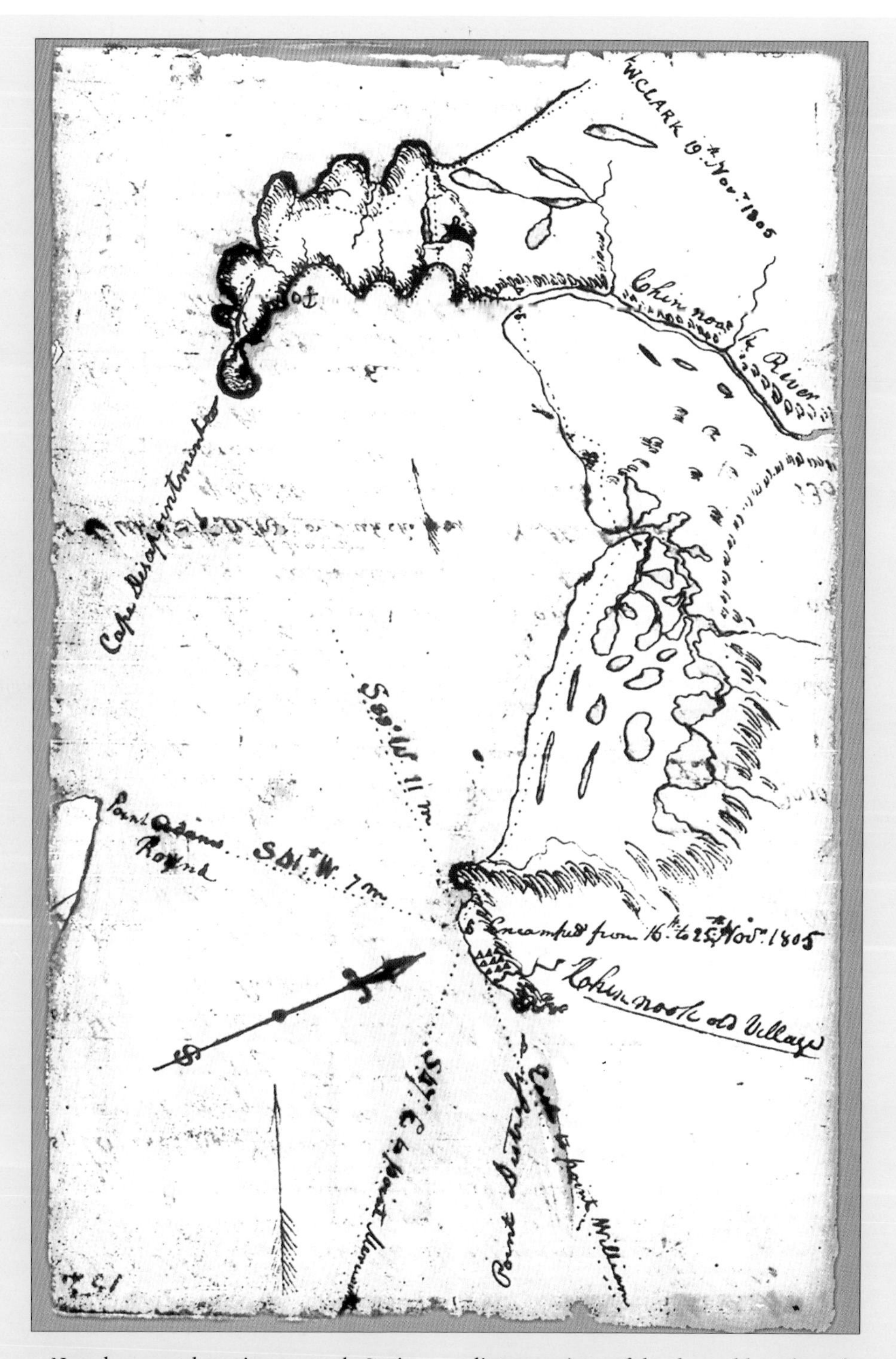

Now, how was he going to mark Station Camp? A wooden stake in the ground would deteriorate; a pile of rock could wash away in the next storm. However, Clark knew that a careful triangulation with his compass from distant points of land would endure for centuries and was absolutely the most reliable way to communicate the location of this historic camp.

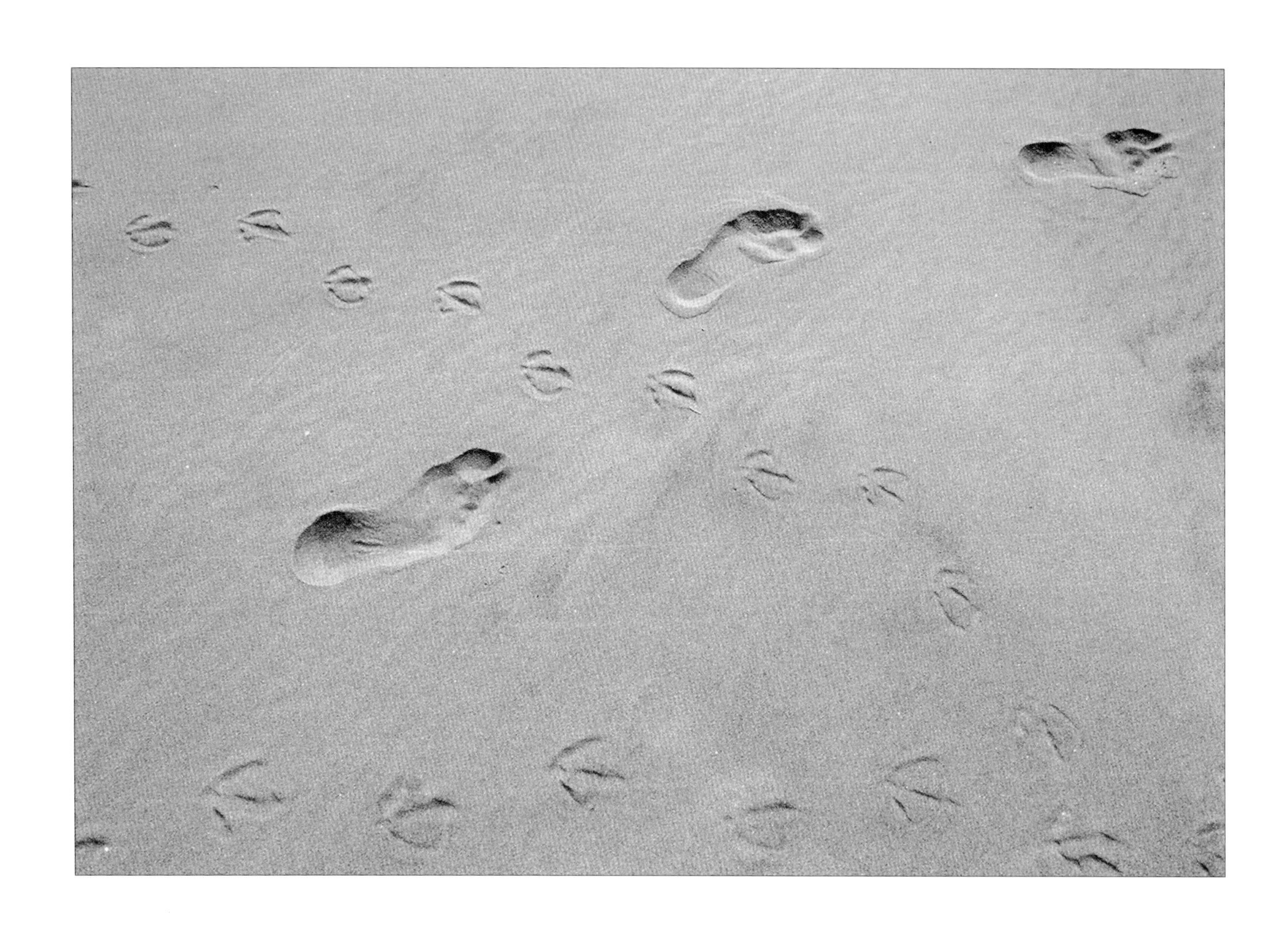

CHAPTER FOUR

Clark's Excursion

the men appeared much satisfied with their trip

Monday, November 18th

Fair weather continues but the jet stream weakens and moves southward. Clouds move in as a low pressure system moves toward the coast.

Daytime High Tide:	*11.02 am*	*9.1*
Daytime Low Tide:	*5:52 pm*	*−0.3*
Sunrise:	*7:19 am*	
Sunset:	*4:45 pm*	

The day began early for Captain Clark and his men. They faced a long and difficult walk, so there was no time to waste.

> *I Set out with 10 men and my man York to the Ocian by land* Moulton, *Journals* 6:65 (Clark)

The party walked down the beach, crossed over an outcropping of rock, and continued along the shore. This route along the water's edge was the shortest and quickest way to Cape Disappointment. Gigantic driftwood logs were strewn along the beach like one continuous wooden trail, and the men probably walked on top of them, rarely setting foot on the ground. Periodically, Clark would stop to measure distant points of land with his compass and write down the bearings. This information was essential for his map.

After several hours of walking, the party arrived at a river too wide and deep to cross. Fortunately, some Chinook women lived nearby and had a canoe. Clark held up two fishing hooks and pointed to the other side; the women immediately understood.

> *here we were Set across all in one Canoe by 2 Squars to each I gav a Small hook* Moulton, *Journals* 6:65 (Clark)

The Indians Lewis and Clark encountered here at the mouth of the Columbia had already been severely impacted by contact with foreigners. Merchants had brought steel tools and guns into this region, but they also had brought fevers, coughs, and venereal disease. Clark observed the plight of these unfortunate natives but was powerless to give any aid.

> *at the Cabin I saw 4 womin and Some Children one of the women in a desperate Situation, covered with Sores Scabs & ulsers no doubt the effects of venereal disorder which Several of this nation which I have Seen appears to have* Moulton, *Journals* 6:65 (Clark)

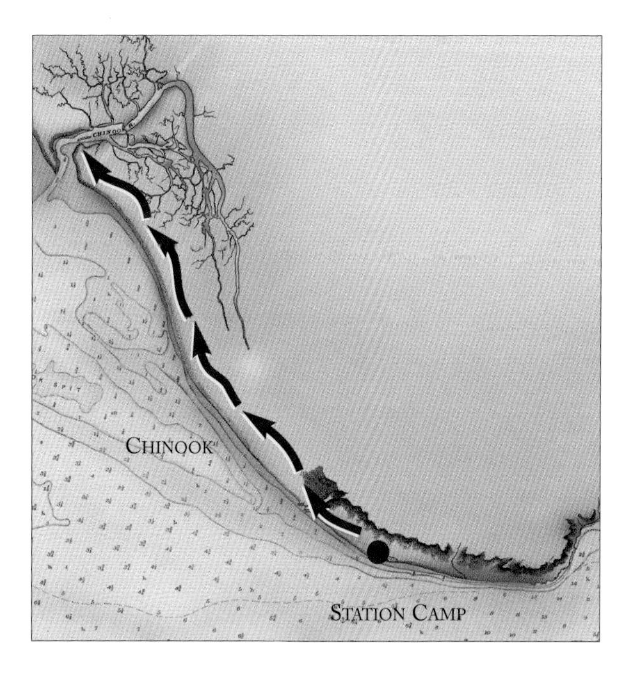

Clark and his party set out early on their excursion to Cape Disappointment. They walk along the shore until they arrive at a large river too deep and wide to cross without a canoe.

Once across the river, Clark's party hurried along their way. They had brought enough food for only one supper, so the men kept close watch for deer or elk. Up ahead along the shore, they saw something curious. It was a large piece of something fleshy, and on top of it there was an odd creature moving around. Being a man of science, Clark wanted a closer look, and there was only one way to do this. He sent Reuben Field on ahead with his rifle.

Both Field brothers were excellent marksmen. When Reuben pulled the trigger, the creature instantly fell dead to the ground. Clark and his men jumped up from their concealment and rushed forward, anxious to discover what it was.

To their astonishment, they found the lifeless creature to be a bird – a bird that was larger than any they had ever seen before. Its body was huge, with a wingspan as wide as a room.

In 1805, the bird that later became known as the California condor covered a range that extended from the Rocky Mountains to the Pacific Ocean. Lewis and Clark killed three condors during their time along the Columbia River.

> *Rubin Fields Killed a Buzzard of the large Kind near the meat of the whale . . . measured from the tips of the wings across 9 1/2 feet*
>
> Moulton, *Journals* 6:66 (Clark)

Confronted with such a specimen, Clark now had to switch from his role of expedition leader to that of zoologist. He knew that President Jefferson would want a full description of this rare and remarkable creature, so he carefully examined it and measured its many different dimensions, right down to a quarter of an inch.

> *from the point of the Bill to the end of the tail 3 feet 10 1/4 inches, middle toe 5 1/2 inches, . . . wing feather 2 1/2 feet long . . . tale feathers 14 1/2 inches, and the head is 6 1/2 inches including the beak*
>
> Moulton, *Journals* 6:66 (Clark)

This bird closely resembled the turkey vultures they were familiar with, but was much, much larger in every conceivable way. They removed its head, which they intended to dry, preserve, and carry all the way back to Philadelphia as a specimen.

The day was slipping past. The party hurried along, crossed another river, and within a mile arrived at a cove. This was the end of the low marshy shoreline where walking had been easy and the beginning of the rugged Cape Disappointment peninsula. From here on their route would be grueling. The steep angle of the terrain, combined with chest-high brush crisscrossed with fallen trees, made their progress slow and difficult. It can take hours to walk even one mile through some of this rugged coastal wilderness.

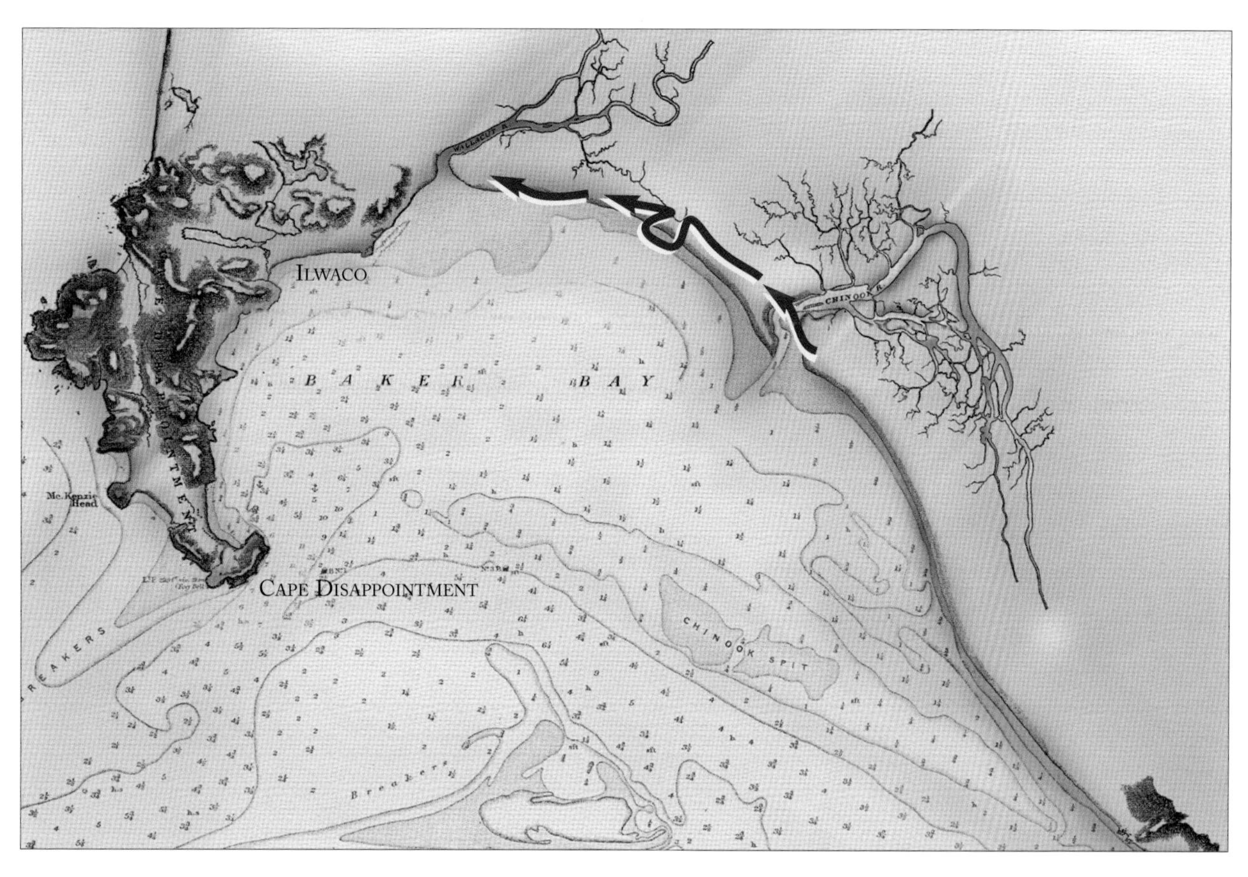

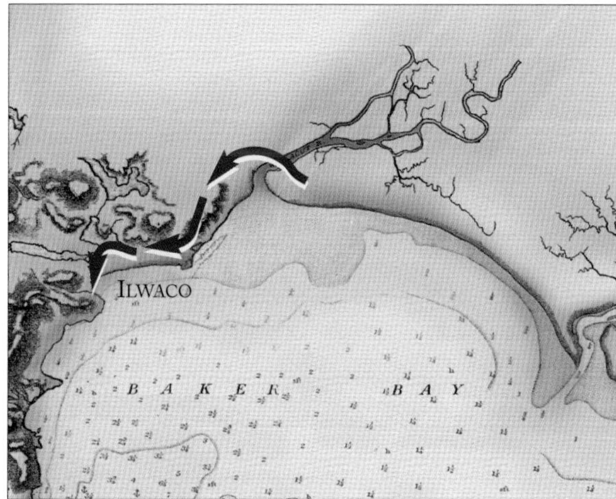

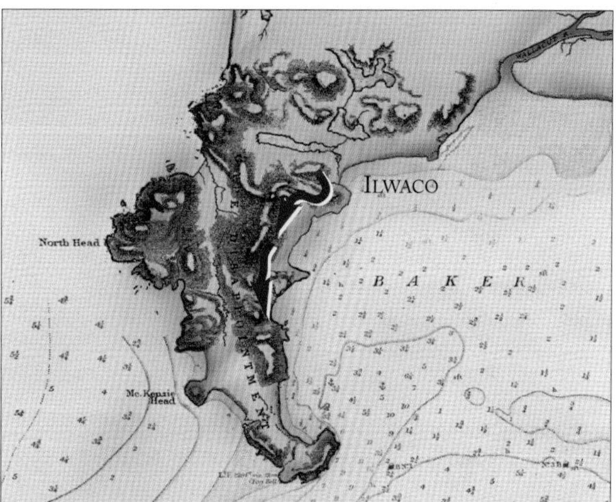

(top) They cross the river and proceed on another mile. Here Reuben Field shoots a California condor. After examining this enormous bird they continue on until they arrive at a second river.

(left) Borrowing a canoe, they cross this second river and continue along the shore until they arrive at the edge of a densely forested cape.

(right) The party scramble along the steep hillside and make their way out towards the tip of the cape. After approximately one mile they come to a cove with a tiny island. This is the anchorage where fur traders and Indians meet to barter.

The millennia of weather and waves had notched three circular coves into the cape. Clark and his men passed from the first cove into the second, and from the second into the third. This third cove, with its distinctive tiny island, was the rendezvous location where American, British, and other European fur traders would anchor their ships while negotiating with local natives.

> *this rock Island is Small and at the South of a deep bend in which the nativs inform us the Ships anchor, and from whence they receive*

The Anchorage

Trade between natives and white men

The coastlines of Washington and Oregon are crowded with steep hills that offer few opportunities for ships to anchor. During the 18th century, ship captains from Spain, France, Russia, and Britain were constantly searching for protected harbors or rivers where they could safely take on fresh water and trade with the natives. The Columbia River was discovered in this way.

In 1775, Spanish Captain Bruno de Heceta claimed he saw a large river near the latitude of 46°; a British captain named Meares attempted to confirm this discovery, but failing to find the river, he expressed his disgust by naming a high rocky point "Cape Disappointment."

The Spanish discovery, however, was not a mistake. In 1792, Captain Robert Gray sailed his ship into the river and found a native population ready and willing to trade. From that moment on, the mouth of the Columbia River and its deep-water anchorage became well known among the Pacific traders.

Ships from Europe and Boston arrived at this tiny cove to trade with the Chinooks. The site was well protected from sudden gales and the water was deep. Lewis described the activity.

> *The traders usually arrive in this quarter . . . in the month of April, and remain untill October; when here they lay at anchor in a bay within Cape Disappointment on the N. side of the river; here they are visited by the natives in their canoes who run along side and barter their comodities with them* [1] (Lewis)

The natives traded hides, furs, fresh food, and sometimes slaves in exchange for cloth, clothing, steel tools, guns, and beads.[2]

> *at 3 miles passed a nitch – this rock Island is Small and at the South of a deep bend in which the nativs inform us the Ships anchor, and from whence they receive their goods* [3] (Clark)

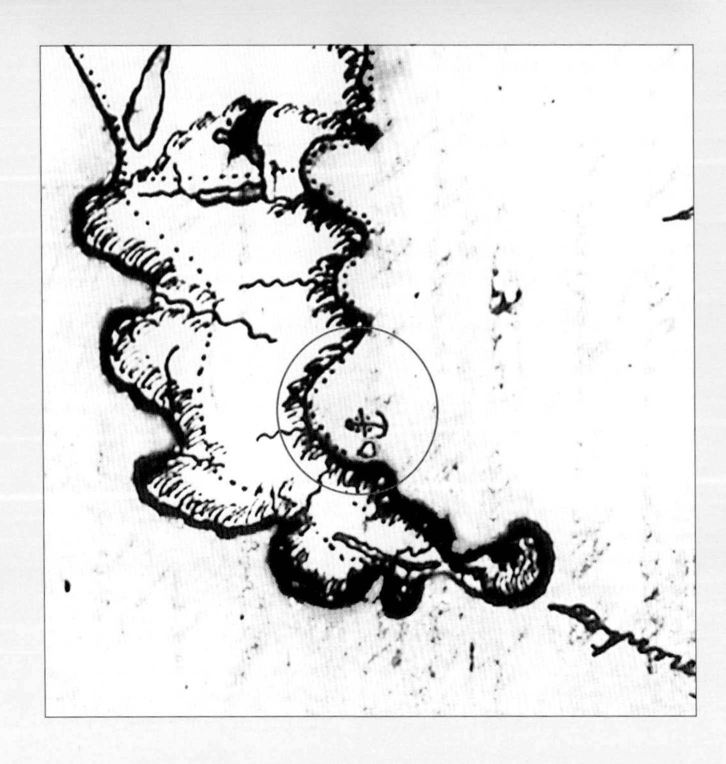

(above) The small island Clark noted at the anchorage still exists and is quite apparent during high tides.

Clark's map of the cape is remarkable for its accuracy. He drew a small ship anchor to mark the cove where ships traded with local Indians.

Clatsop and Chinook Indians provided Lewis and Clark with the names of traders, the type of ship they sailed, and when they expected them to return. They were so specific with their details, they even noted that one man had a wooden leg and another had only one eye.

their goods in return for their peltries and Elk Skins &c. this appears to be a very good harber for large Ships

Moulton, *Journals* 6:66 (Clark)

Graffiti were everywhere. Throughout the previous dozen years, fur traders had made a point of carving their names, the name of their ship, and the date of their arrival into the trunks of trees.[1] Lewis had marked one tree with his name when he passed by, and Clark decided to do the same. In order to make sure his name wouldn't be confused with that of some ship captain, Clark included the words "by land." It was subtle, but enough to inform anyone who saw it that someone had arrived here the hard way.

here I found Capt Lewis name on a tree. I also engraved my name & by land the day of the month and year, as also Several of the men

Moulton, *Journals* 6:66 (Clark)

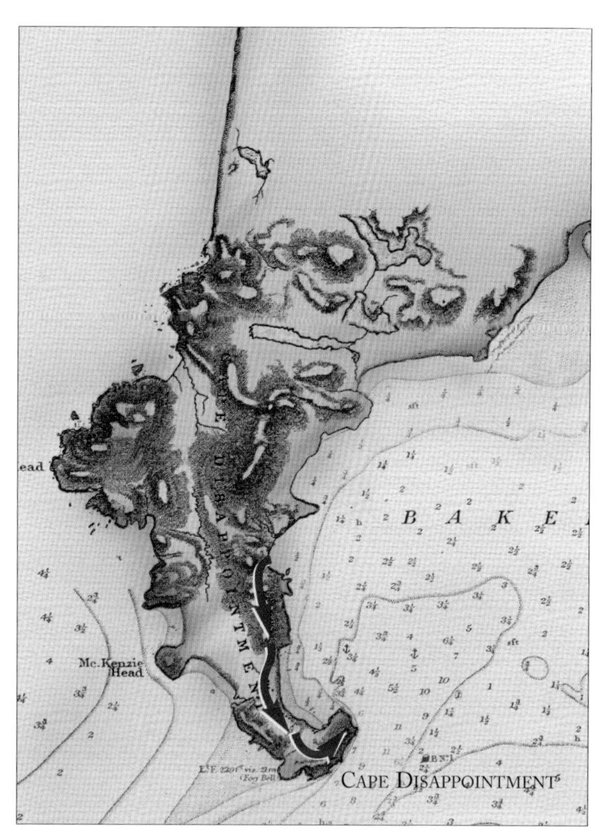

Clark proceeds to the inner tip of the cape and makes a detailed set of observations with his compass to distant landmarks.

From this cove Clark's route is not well defined; he may have followed either of two different routes. From his map it appears that he turned and crossed over to the ocean side of the cape. However, his journal tells a different story. It states that he continued on to the extreme tip of the cape.

to the iner extremity of Cape Disapointment passing a nitch in which there is a Small rock island

Moulton, *Journals* 6:66 (Clark)

After he finished making his last set of compass readings, Clark backtracked and then crossed the cape to the ocean side. This route, as described in his journal, seems logical and is consistent with his effort to measure the shoreline.

from the last mentioned nitch . . . I crossed the neck of Land low and 1/2 of a mile wide to the main Ocian

Moulton, *Journals* 6:66 (Clark)

The party backtracks a short distance and crosses over to the ocean side of the cape. They find a large mound of solid rock surrounded on three sides by ocean. It provides an excellent place to view the vast Pacific.

Years later, the large rock mound which Clark and his party ascended was named "McKenzie Head" and is plainly visible in this 1905 photograph, looking south towards Cape Disappointment.

Once they reached the ocean side of the cape, Clark and his party found themselves standing along a short, sandy beach with steep cliffs bordering either side. A large, hump-backed mound of rock towered above the crashing surf directly in front of them. One narrow edge touched the beach, giving access to the rock's rounded top. Without hesitation, Clark and his men scrambled up for a commanding view of this vast ocean.

> *at the foot of a high open hill projecting into the ocian, and about one mile in Sicumfrance. I assended this hill which is covered with high corse grass*
> Moulton, *Journals* 6:66-67 (Clark)

The view was spectacular. Directly in front of them was the Pacific Ocean, the largest body of water on the planet. To their left and right, the ocean's surf relentlessly pounded against the cliffs of rock. Sea foam and spray tossed high into the air; the thumping of each wave reverberated with a deep, explosive roar that sounded like an underground thunderstorm. When they licked their lips the men tasted a bitter trace of salt.

> *men appear much Satisfied with their trip beholding with estonishment the high waves dashing against the rocks & this emence ocian*
> Moulton, *Journals* 6:67 (Clark)

Evening was rapidly approaching, so after a brief pause to admire the ocean from this elevated view, they descended to the northern side of the rock and set up camp. Driftwood was strewn everywhere, and the men found it easy to curl up alongside the logs, perfectly protected from the wind. A campfire would have been blazing within minutes.

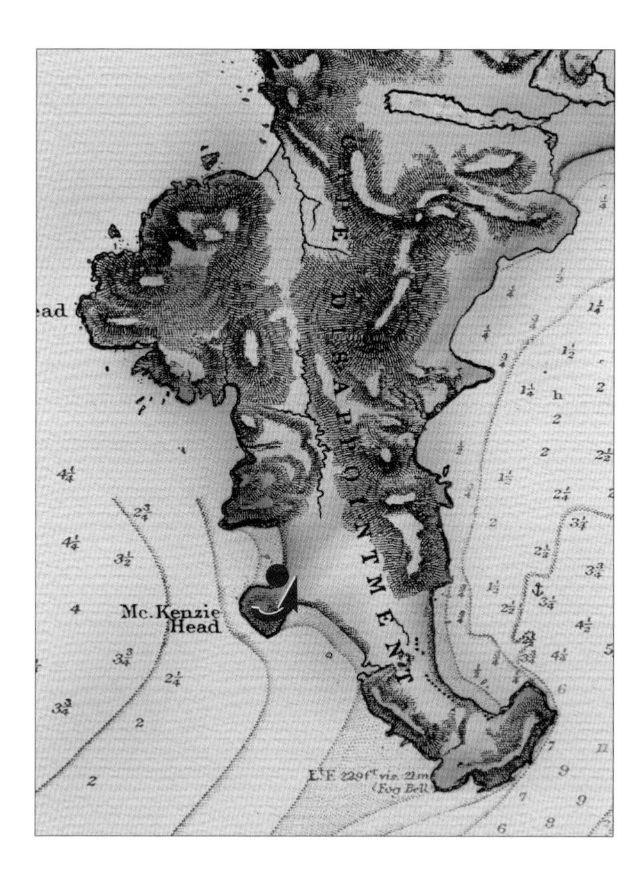

decended to the N. of it and camped

Moulton, *Journals* 6:67 (Clark)

Captain Clark and his men had walked more than a dozen miles this day and must certainly have felt the fatigue from such exertion. As darkness came they laid out their blankets.

The conversations that evening between Clark and his men as they lay around their campfire could have covered many subjects. Perhaps the men remarked how peculiar it had been to walk through miles of perfect deer and elk habitat without seeing a single animal. Or maybe they talked about the huge vulture eating the whale, or the names they saw carved into trees. It is very likely they talked about such things, but this is only speculative. The only conversation between Clark and his men of which we know for sure involved a discussion about their upcoming winter campsite. Now that their westward journey was finished, it was time to begin thinking about their return. Apparently Clark questioned his men about their preferred winter campsite.

> *men all Chearfull, express a Desire to winter near the falls this winter* Moulton, *Journals* 6:65 (Clark)

There were two waterfalls on the Columbia River, and this entry could refer to either one;[2] but it seems most likely that the men were showing an inclination towards the falls above the Long Narrows. The Indians there were friendly, they had abundant dried food, and it was close to the foothills where they had left their horses among the Nez Perce. It would make a fine winter camp.

There was no sunset. The sun simply disappeared behind a thick, gray wall of clouds that hung like a curtain across a stage. Dry, calm days do not last long here in November, and this band of clouds over the ocean was the first sign that the weather was beginning to return to its more typical mood.

Clark's party descends McKenzie Head, walks a short distance towards the north, and camps.

Meanwhile, back at camp, Lewis passed a restful, uneventful day. The hunters brought in more game while other members of the party mended clothes and fixed their worn-out equipment. A handful of curious Indians had set up camp a short distance away, and this gave Lewis an opportunity for more ethnological work. This time he delved into their language.

President Jefferson had given Lewis a list of words he wanted translated into Indian languages. Jefferson had been collecting native languages since the 1770s as part of a comprehensive study of North American linguistics. The vocabulary list contained 364 words, so this task probably consumed several hours of Lewis's day. (*see Appendix Six: Jefferson's Vocabulary List p. 200.*)

Tuesday, November 19th

Rain showers begin; the sky is overcast but winds are calm. A low pressure system pushes the jet stream south of the region.

Daytime Low Tide:	*5:41 am*	*2.3*
Daytime High Tide:	*11:42 pm*	*9.2*
Sunrise:	*7:20 am*	
Sunset:	*4:44 pm*	

The following morning Clark and his men awoke soaking wet.

> *I arose early this morning from under a wet blanket caused by a Shower of rain which fell in the latter part of the last night* Moulton, *Journals* 6.69 (Clark)

Obviously, this would have been a perfect time to turn around and retrace their steps back to the main camp. Instead, Clark turned towards the north, apparently determined to explore the entire circumference of this unusual cape.

Their food was gone, so Clark selected two of his best hunters and sent them ahead. He kept everyone else with him for the time being, so they wouldn't spook the game before the hunters had a chance to approach it.

> *Sent two men on a head with directions to proceed on near the Sea Coast and Kill Something for brackfast . . . that I Should follow my Self in about half an hour* Moulton, *Journals* 6:69 (Clark)

After the hunters had a sufficient head start, Clark and his men followed.

The terrain they were crossing was among the most rugged they had ever experienced. Millions of years of harsh weather had worn the outer edge of the cape into a zigzag of razor-back ridges and deep gullies. The thick forest concealed this from view, so Clark's men had no idea how difficult their hike would be until they were in the middle of it.

Clark and his men were such superb trackers that they could see the vague, nearly invisible, signs left by the hunters. A small circle of compressed moss showed where a foot had recently stepped; missing dewdrops indicated where someone had brushed against a limb. Mile after

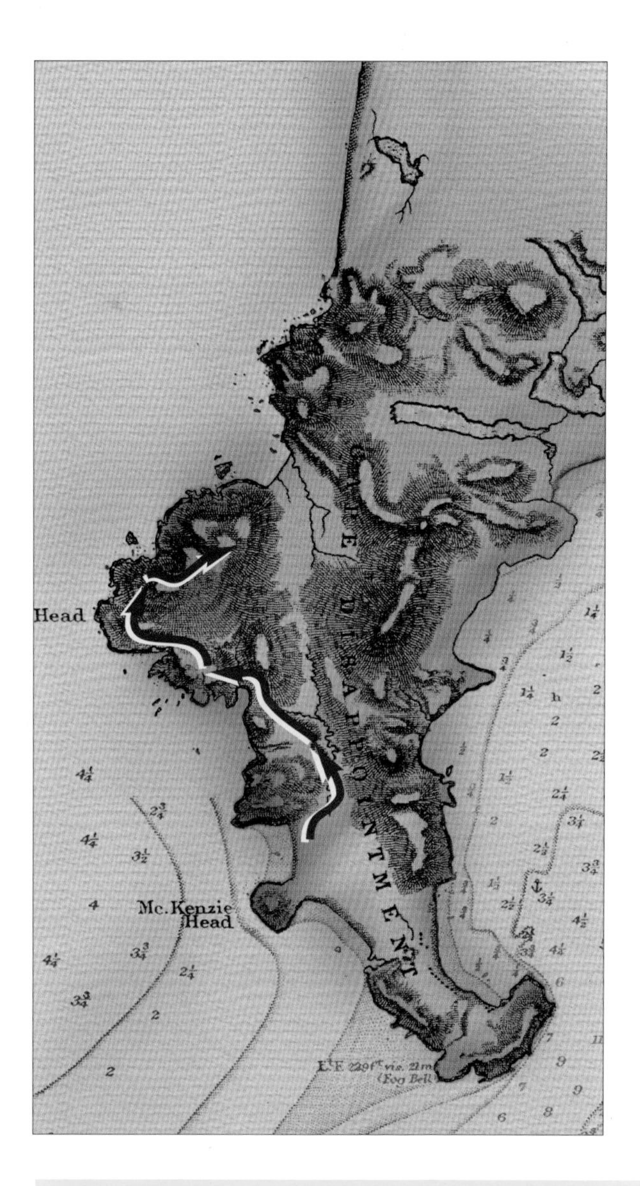

Clark sends out his hunters to shoot something for their breakfast, then follows with the rest of his party.

Tracking

Clark and his men were excellent trackers; it was an essential skill for everyone who traveled in wilderness areas. Here is Lewis's description of Clark discovering an Indian footprint while they were ascending the Missouri River.

> *Capt. Clark set out this morning as usual. he walked on shore a small distance this morning and killed a deer. In the course of his walk he saw a track which he supposed to be that of an Indian from the circumstance of the large toes turning inward. he pursued the track and found that the person had ascended a point of a hill from which his camp of the last evening was visible*[1]
>
> (Lewis)

(above) Spruce trees like Doric columns.

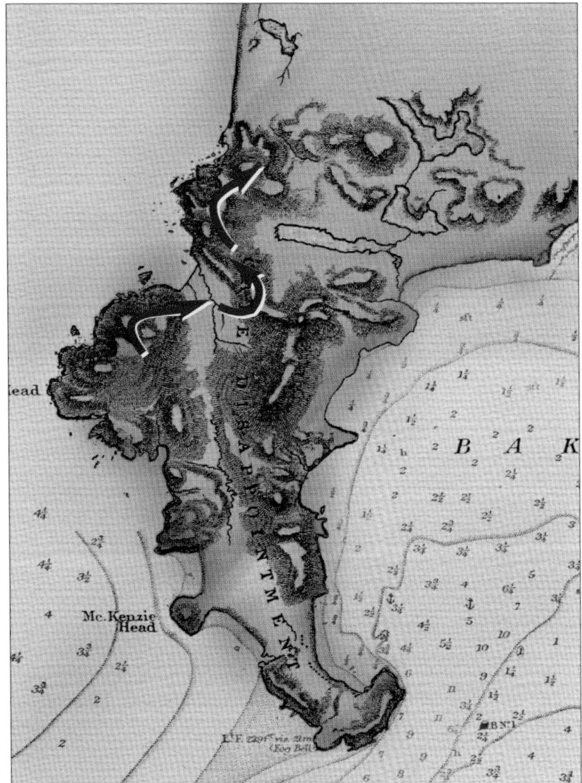

(left) Following their meal of roasted deer, the men continue northward through the rugged cape until they arrive at a rocky bluff that towers above a long, sandy beach.

After a brief inspection of the surrounding countryside, Clark and his men descend the rock cliff and begin a four-mile walk along the coast.

mile they followed along the hunters' route, scrambling up and down the steep and brushy hillside. At last they caught up with their hunters and were pleased to see their success.

> *I overtook the hunters at about 3 miles, they had killed a Small Deer on which we brackfast*
>
> Moulton, *Journals* 6:69 (Clark)

The deer was roasted and consumed by Clark's hungry men. Then, without delay, they continued on northward.

> *after takeing a Sumptious brackfast of venison which was rosted on Stiks exposed to the fire, I proceeded on through ruged Country of high hills and Steep hollers* Moulton, *Journals* 6:69 (Clark)

Their route continued to be challenging. In addition to the uneven ground, they also had to deal with fallen timber and thick patches of salal brush that grew here six feet high. Every step was an effort. The only redeeming aspect of this walk was the beauty of the magnificent Sitka spruce forest, which towered two hundred feet high overhead. These marvelous trees grew straight with barely a taper, like gigantic Doric columns of ancient Greece. Some of these trees had circumferences of thirty to nearly forty feet. It was the type of lush, green forest that would take your breath away.

After several more miles of difficult walking, they reached a point where they were forced to stop. Here the high bluffs and forest dropped straight down to a long gray sand beach. Clark traced this ribbon of sand with his eyes as far as he could see, carefully calculating mile after mile to estimate its length. After a brief examination of the surrounding countryside, the party descended to the beach and continued northward.

For these travelers who had come such a long distance over such rugged and wild terrain, stepping out onto this beach must have felt like a dream. The sweeping motion of the waves smooths the sand as level as a dance floor. For the first time in years, maybe for the first time in their lives, these men could walk without tripping over roots or twisting their ankles in a shallow hole. There were no rocks poking into the bottoms of their feet, no sharp, spiny cactus or poisonous rattle snakes. Walking on this beach may have been the most carefree and enjoyable stroll they had ever experienced. Perhaps some of the men stripped off their moccasins and felt the smooth, soft sand beneath their weary feet.[3]

The men walked along and watched the curling surf rise up and break onto the beach with a thunderous roar. Strewn everywhere was driftwood, polished as bright as silver by windblown sand. Sometimes this driftwood was

Walking

The feet of Lewis and Clark's men took a terrible beating during this expedition. While crossing western Montana Lewis wrote:

> *the sharp points of earth as hard as frozen ground stand up in such abundance that there is no avoiding them. this is particulary severe on the feet of the men... some are limping from the soreness of their feet, others faint and unable to stand for a few minutes* [1]

Willapa Bay

Clark said he saw some ponds to the northeast of Cape Disappointment and did not bother to investigate any further. However, if he had, he would have stumbled upon the second largest bay on the west coast. Willapa Bay (known also as Shoalwater) measures one hundred square miles in water area with two hundred miles of shoreline.

> *a low pondey countrey, maney places open with small ponds in which there is great numbr. of fowl I am informed that the Chinnook Nation inhabit this low countrey and live in large wood houses on a river which passes through this bottom Parrilal to the Sea coast* [1] (Clark)

Willapa Bay and Baby Island.

sculpted by nature into fantastic forms that might have appeared like imaginary creatures. The misshapen stump with twisted roots might resemble a horse's head with wind-tossed mane; a peculiarly curved limb might appear like a coiled snake, mouth open, ready to strike. Clark's men were surrounded by natural wonders. They picked up seashells and gathered pebbles of pumice. Meanwhile, Clark jotted down brief descriptions of the things they saw.

> *I saw a Sturgeon which had been thrown on Shore and left by the tide 10 feet in length, and Several joints of the back bone of a whale which must have foundered on this part of the Coast*
>
> Moulton, *Journals* 6:70 (Clark)

After nearly an hour, Clark saw there was no change in the beach. Each mile resembled the previous one. The day was quickly slipping by. A prominent pine tree standing near the shore made an ideal place to turn around. Clark stopped there long enough to carve his name.

> *I proceeded on the Sandy Coast 4 miles, and marked my name on a Small pine, the Day of the month & year*
>
> Moulton, *Journals* 6:70 (Clark)

Clark and his men now turned southward and retraced their steps. They could plainly see the rugged, broken cliffs of Cape Disappointment looming in front of them. No one was looking forward to revisiting that

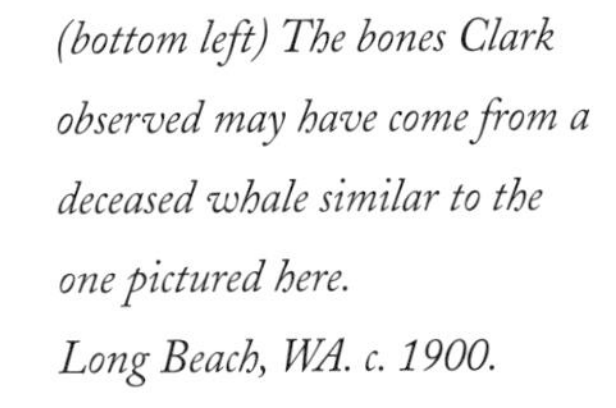

(bottom left) The bones Clark observed may have come from a deceased whale similar to the one pictured here.
Long Beach, WA. c. 1900.

steep, dreadful place. Fortunately for them, Clark was a gifted cartographer. He could look at his compass while walking and draw a precise picture of the landscape in his mind. From this mental picture of the terrain he could see exactly where he was and where he was trying to go.

The narrow shape of this cape, which resembled an icicle, provided an opportunity to avoid all the uneven terrain.[4] All they had to do was to cut across a couple of miles to the southeast and they would arrive directly back at the first cove. This shortcut would save them nine miles of hard walking, hours of time, and a tremendous amount of energy.

> *returned to the foot of the hill, from which place I intended to Strike across to the Bay . . . I proceeded through over a land S E with Some Ponds to The bay distance about 2 miles* Moulton, *Journals* 6:70 (Clark)

Precisely as he had planned, Clark and his party arrived back at the shore of the Columbia. From this cove they backtracked along the Columbia and camped near the mouth of the river they had crossed the day before.

> *thence up to the mouth of Chinnook river 2 miles, crossed this little river in the Canoe we left at its mouth and Encamped on the upper Side in an open Sandy bottom* Moulton, *Journals* 6:70 (Clark)

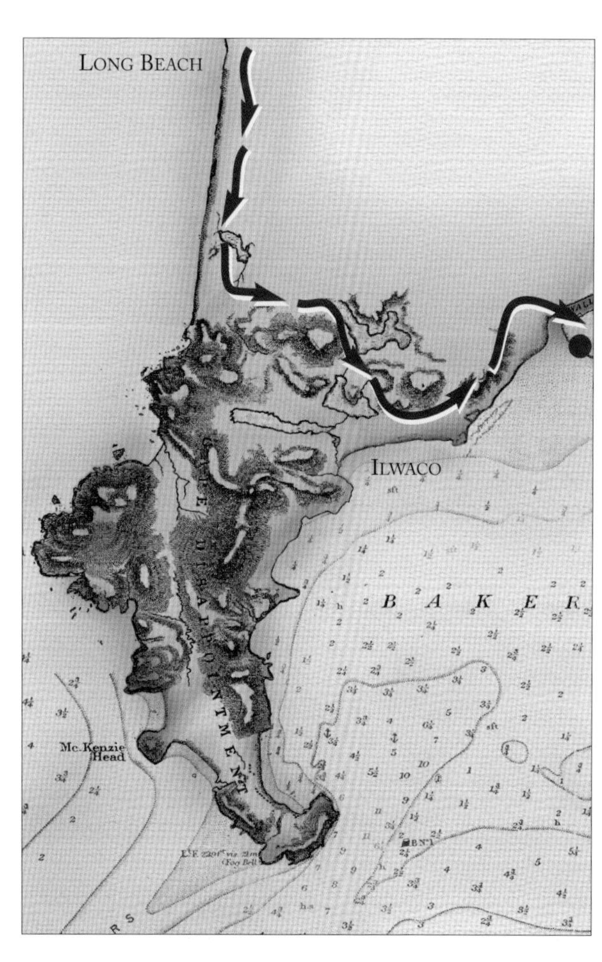

(left) This sturgeon measured slightly longer than the one Clark described. Caught in the Columbia River in 1916, this creature weighed about 750 pounds.

(right) Clark carves his name in a tree, turns around, and leads his party back down the beach towards the cape. Rather than retrace their route through the rugged highlands, Clark calculates a shortcut to the Columbia. His estimation is perfect and the party arrives exactly where he intended. They continue upriver along the shore and camp for the night.

This second day of Clark's excursion ended quietly and without food. There was no mention of dinner. The deer meat was gone and nothing else had been shot. Nevertheless, Lewis and Clark's men were having many new and marvelous encounters here near the ocean, but who back home would ever believe such fantastic stories? How would they describe a gigantic bird eating a whale? Who would ever believe a fish that measured ten feet? Would anyone believe a story of one rainstorm that lasted ten days? How would they describe trees thirty-five feet in circumference that grew so tall that their tops disappeared into the clouds? Upon hearing these stories, their families and friends would probably laugh and ask them, "Where do you think you have been – to the land of the giants?"

Captain Lewis seems to have had another quiet day back at the main camp. Various Indians came to visit dressed in fabulous clothes, presumably to impress these visitors. Joseph Whitehouse described the scene:

> *A number of Indians came to visit us at our Camp. They wore Robes made out of the Skins of swans, Squirrel skins, & some made out of beaver skins also . . . These Indians are a handsome well looking set of People, and were far the lightest colour'd Natives that we had seen since we have been on our Voyage*
>
> Moulton, *Journals* 11:395-396 (Whitehouse)

Where are the Trees?

Lewis and Clark were astounded at the size of the trees that grew in this coastal rainforest. They repeatedly measured and described them in their journals.

the most common species . . . in this neighbourhood. it appears to be of the spruse kind. it rises to the hight of 160 to 180 feet very commonly and is from 4 to 6 feet in diameter[1] (Lewis)

The Pine of fur Specs, or Spruc Pine grow here to an emense Size & hight maney of them 7 & 8 feet through and upwards of 200 feet high[2] (Clark)

The hills next to the bay Cape disapointment to a Short distance up the Chinnook river . . . thickly Coverd. with different Species of pine &c. maney of which are large, I observed in maney places pine of 3 or 4 feet through growing on the bodies of large trees which had fallen down, and covered with moss and yet part Sound[3] (Clark)

There are sveral species of fir in this neighbourhood . . . which grows to immence size; very commonly 27 feet in the girth six feet above the surface of the earth, and in several instances we have found them as much as 36 feet in the girth or 12 feet diameter . . . they frequently rise to the hight of 230 feet[4] (Lewis)

The pioneers who settled this area also admired these trees, especially when they discovered the wood from these forests was of a quality superior to anything they had ever seen. They cut these trees, milled the wood into lumber, and built their houses, docks, fishing boats, railroad ties, boardwalks, canneries, and more sawmills. The cutting of these forests on private, state, and federal land continued throughout the twentieth century; only a few old trees remain today.

Wednesday, November 20th

Rain showers continue; the sky is overcast but winds are calm. Out in the Pacific Ocean a low pressure system intensifies.

Daytime Low Tide:	*6:25 am*	*2.6*
Daytime High Tide:	*12:20 pm*	*9.1*
Sunrise:	*7:20 am*	
Sunset:	*4:44 pm*	

THE THIRD DAY OF CLARK'S EXCURSION BEGAN much like the day before. Once again they awoke soaking wet and hungry. Hunters were immediately sent out while the others dried themselves by the fire.

> *Some rain last night despatchd. 3 men to hunt Jo. Fields & Colter to hunt Elk & Labich to kill some Brant for our brackfast* Moulton, *Journals* 6:71 (Clark)

Several hours later, after a breakfast of roasted duck, Clark and his men set out; however, when they arrived at the river, their travels came to an abrupt stop. The Chinook women who had ferried them across two days earlier were gone, and all the canoes were on the opposite shore.

> *I proceeded on to the enterance of a Creek near a Cabin no person being at this cabin and 2 Canoes laying on the opposit Shore from us*
>
> Moulton, *Journals* 6:72 (Clark)

Clark had two choices: either wait for the return of the Chinook women, or somehow get his men across without them. He chose not to wait. The men flew to work with their tomahawks; logs were chopped into manageable lengths. Clark's journal doesn't tell how long this took, nor does he tell us how they managed to lash the logs together; but somehow these remarkably skillful woodsmen constructed a raft.

> *a Small raft was Soon made, and Reuben Fields Crossed and brought over a Canoe*
>
> Moulton, *Journals* 6:72 (Clark)

Now there were no more obstacles. The men took advantage of the ebbing tide and walked along the narrow shoreline, making good time on their return to camp.

> *The tide being out we walked home on the beech*
>
> Moulton, *Journals* 6:72 (Clark)

They did, however, encounter several parties of Indians, who all appeared to recognize Clark.

> *on my way up I met Several parties of Chinnooks which I had not before Seen* Moulton, *Journals* 6:72 (Clark)

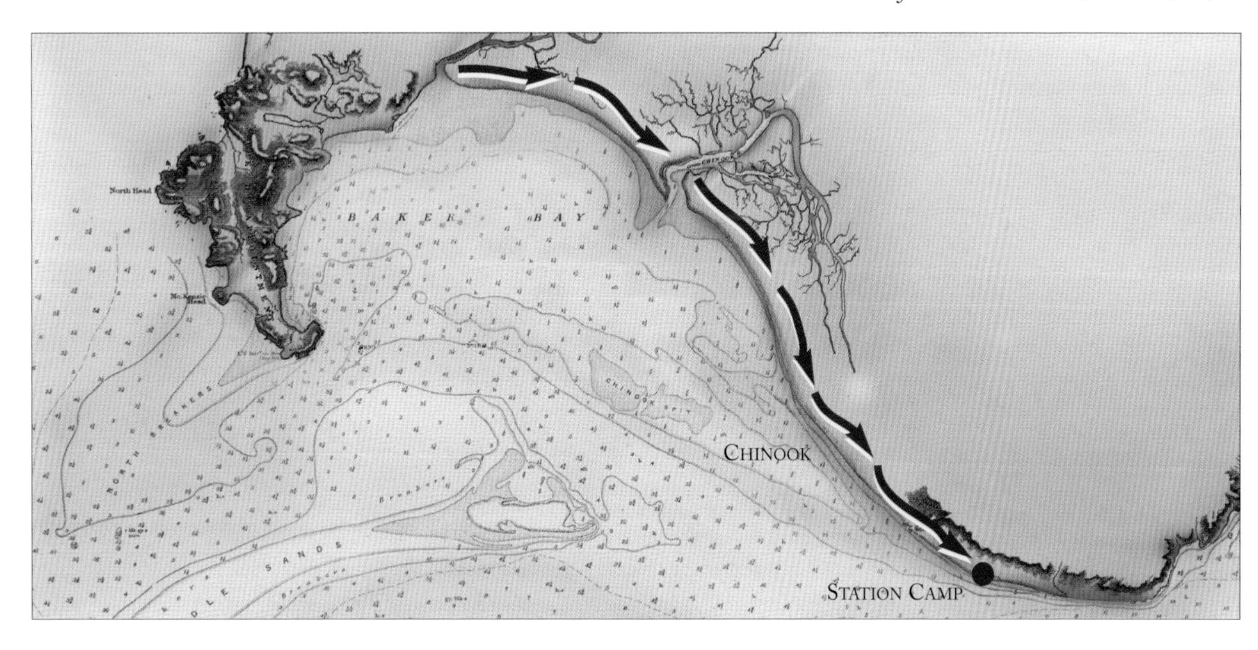

Clark and his men proceed along the shore of the Columbia a couple of miles until they come, once again, to the broad river that blocks their route. The men assemble a raft, cross the river, and are on the move once again; they reach Station Camp within a few hours.

Clark's threat to kill any Chinook caught stealing had spread rapidly throughout the Indian community. The Chinooks knew what had been said and which one of these men had made the threat. They immediately spotted Clark and shied away from him. We can easily imagine them saying to one another as Clark approached, "There he is – stay away from that one – he's trigger-happy – that's the one who said he would have us shot!"

> *all those people appear'd to know my deturmonation of keeping every individual of their nation at a proper distance, as they were guarded and resurved in my presence* Moulton, *Journals* 6:72 (Clark)

When Clark finally arrived at Station Camp, he found several Chinooks had come to trade. One of them had brought a fur robe to sell, and upon close examination Clark realized it was made of sea otter.

Sea otter was the most highly prized fur in the world, far outclassing that of any other animal. Clark was completely smitten by its extraordinary beauty.

> *one of the Indians had on a roab made of 2 Sea Otter Skins the fur of them were more butifull than any fur I had ever Seen* Moulton, *Journals* 6:72 (Clark)

Lewis and Clark desired to purchase this fur, perhaps to add to their collection of curiosities intended for President Jefferson. They unpacked some trade goods and began negotiating.

> *both Capt. Lewis & my Self endeavored to purchase the roab with differant articles*
> Moulton, *Journals* 6:72-73 (Clark)

They offered a variety of items, but none of them were valuable enough to entice the Indian to sell.

> *Capt. Lewis offered him many things for his Skins with others a blanket, a coat all of which he refused*
> Moulton, *Journals* 6:72 (Clark)

Fortunately, Sacagawea happened to be wearing a belt that apparently caught the attention of this Chinook man. It was made of blue beads, the one thing the Northwest Indians could not resist. Lewis and Clark recognized this opportunity, quickly offered up the belt, and the Chinook agreed to the trade.

> *at length we precured it for a belt of blue beeds which the Squar – wife of our interpreter Shabono wore around her waste* Moulton, *Journals* 6:73 (Clark)

We have no way of knowing whether Sacagawea was immediately offered anything in exchange for her belt, nor do we know how eager she was to part with it. All we know for sure is that the Chinook ended up with her belt, and Lewis and Clark got the sea otter robe.

The remaining hours of this day passed uneventfully; the pages of Clark's journal mention nothing out of the ordinary. However, looking back, we can identify this as a very significant day. It was the pivotal point in the expedition, the moment when Lewis and Clark finally turned their gaze for the first time away from the west and began to plan their homeward journey.

A brief review of their situation will help clarify why they intended to leave the ocean and head back upriver.

1. First and foremost, their mission had been a success. They had arrived at the Pacific Ocean, exactly as President Jefferson had instructed (*see Appendix Four, Jefferson's Letter to Lewis, 1803, p. 196*).

Four possible winter campsites observed by Lewis and Clark during their descent of the Columbia River, October and November, 1805.

On the night of November 18, while encamped near Cape Disappointment, Clark's men indicated that they preferred making a winter camp several hundred miles upriver near the Celilo Falls. They had been well received by the natives there and noticed they had plenty of surplus food.

The nativs of this village reived me verry kindly, one of whome envited me into his house, which I found to be large and comodious[1] (Clark)

I counted 107 Stacks of dried pounded fish in different places on those rocks which must have contained 10,000 w. of neet fish[2] (Clark)

The men quickly formed warm relations with these Indians.

Crusat played on the violin and the men danced which delighted the nativs, who Shew every civility towards us[3] (Clark)

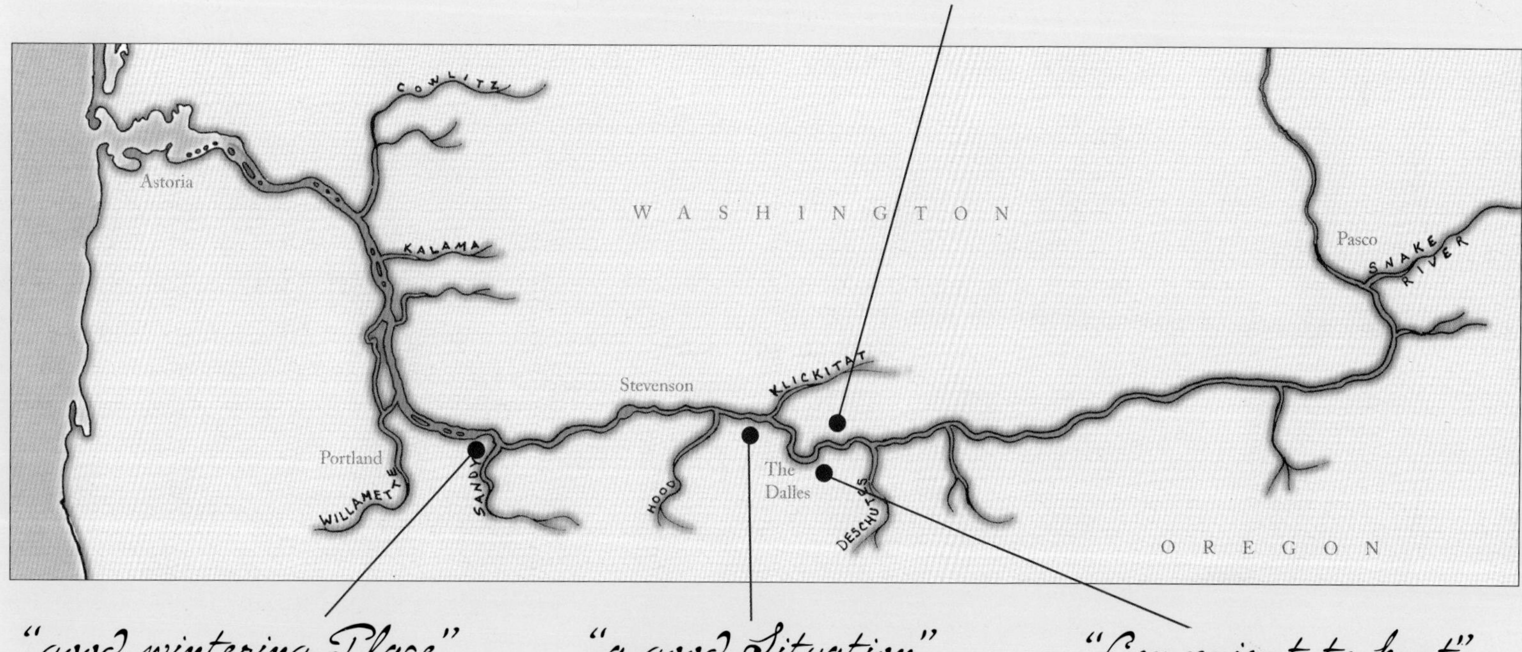

"good wintering Place"

While descending the Columbia River on November 3, Lewis and Clark recognized an excellent winter campsite near the mouth of the Sandy River. There was plenty of timber for the construction of houses, the climate appeared mild, and deer, elk, bear, ducks, and geese were abundant.

opposit qk Sand River... extensive bottoms and low hilley Countrey on each Side (good wintering Place)[4] (Clark)

"a good Situation"

On October 29, the party was among Indians who treated them kindly and provided them with delicious food. Clark writes:

Those people are friendly gave us to eate fish beries, nuts bread of roots & Drid Beries and we Call this the friendly Village[5]

On the opposite side of the river was a campsite that appeared to be defensible and had access to good hunting. Fresh water and timber were nearby.

a good Situation for winter quarters if game can be had is just below Sepulchar rock[6]

"Conveniant to hunt"

At the site of present-day The Dalles, Oregon, Lewis and Clark found a fine campsite on top of a large, flat rock, with near-vertical sides. Food was available a short distance away.

we formed our Camp on the top of a high point of rocks, which forms a kind of fortification... this Situation we Concieve well Calculated for defence, and Conveniant to hunt under the foots of the mountain to the West... where timber of different kinds grows, and appears to be handsom Coverts for the Deer[7] (Clark)

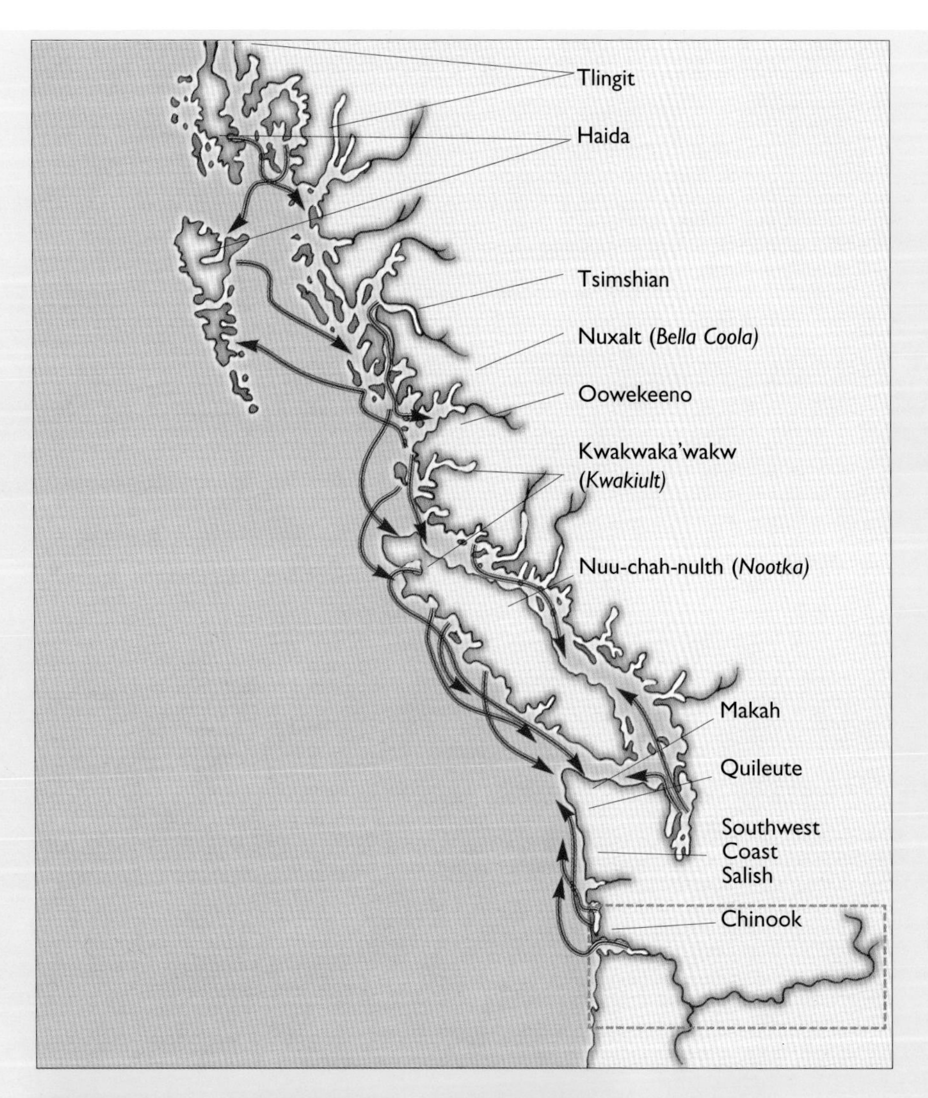

The Network of Trade

The Northwest tribes were culturally advanced, sophisticated, and enjoyed highly developed weaving, basketry, sculpture, and woodworking. Hundreds of thousands of people, living in scattered villages and speaking thirty-nine languages, inhabited this vast region of islands, bays, inlets, and harbors.

Commodities were traded up and down the coast. Some served practical purposes whereas others were purely ornamental. From the tribes in the north came ceremonial robes made of mountain-goat hair, copper plates, spruce-root baskets, dugout cedar canoes, mountain-goat horn spoons, dentalia shells, wooden spoons, oolanchen (candlefish), dog-hair blankets, and woven cedar-bark capes.

Far to the south and two hundred miles up the Columbia River, a series of narrow rapids and waterfalls forced migrating salmon to delay their struggle upriver, and this allowed nearby Indians all the time they needed to catch huge quantities of delicious fish. The surplus food attracted people from miles around who brought items from their homelands for barter. Over time, this area became the largest trade center in the Northwest.

Items such as mountain sheep horns, buffalo blankets, obsidian, jade, native tobacco, bear grass, camas roots, horses, baskets and slaves were brought to trade for salmon. Many of these items were traded downriver to the Chinooks and later sold to northern tribes.

The Chinook occupied a midway point between the upriver market and the populous Northwest Coast. Large

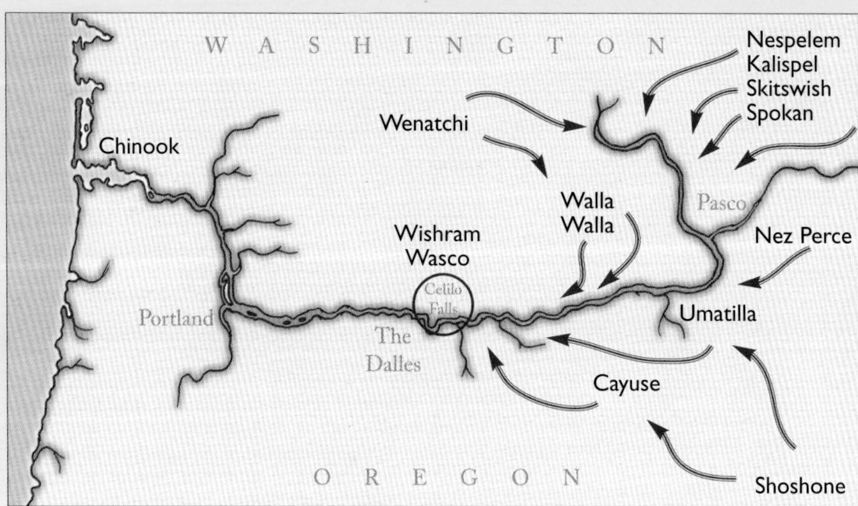

quantities of merchandise, both in raw materials and finished products, passed through their hands as it was shipped up and down the Columbia. Possessing this vital link between river and ocean allowed the Chinook to create tremendous wealth and influence.

Long slender sea shells from Vancouver Island were traded up the Columbia River, into the Rocky Mountains and Central Plateau. Trade items included basketry as well as spoons and bowls made of horn. Net anchors are all that remain of the great salmon fisheries near The Dalles, Oregon, were this trade occurred.

2. There were no ships from which they could replenish their supplies.[5]

3. Food was scarce. In the past thirteen days they had shot dozens of geese, but fewer than twenty deer. Large game animals, such as elk or bear, apparently did not live near the coast.[6]

4. Chinook prices were unaffordable. Everything near the ocean sold at inflated prices intended for dealing with crews from sailing ships. Clark calculated that they would need ten times as many trade items as they possessed in order to survive just one winter among the Chinook.[7]

The combination of these facts was more than enough to persuade Lewis and Clark that there was no reason to remain at Station Camp any longer. There would be better hunting upriver, and the Indians there would sell their surplus food at more affordable prices. So it was decided: tomorrow they would break camp, load their canoes, say good-bye to the ocean, and head upriver. Tomorrow would begin their journey home.

As darkness fell over the mouth of the Columbia, the men retired for one last time into their little wooden shelters. They were expecting many days of hard paddling, and this would probably be their last comfortable night for the next several weeks.

Unfortunately, six hundred miles out in the ocean and heading directly towards them was a tremendous winter gale. There was no sign of it yet, but the weather would soon be changing.

CHAPTER FIVE

An Unexpected Change of Plans

our homeward bound journey

Thursday, November 21st

The approaching storm affects the coastal weather. Rain falls in hard showers without a pause and the wind increases hour by hour.

Daytime Low Tide:	*7:07 am*	*2.9*
Daytime High Tide:	*12:56 pm*	*9.0*
Sunrise:	*7:22 am*	
Sunset:	*4:42 pm*	

Curious local Indians had come to visit and trade with these white men. Some had remained all night near Station Camp. On this morning, however, the Indians looked at the sky, immediately jumped in their canoes and took off. Overhead the coastal birds likewise were heading inland to protected ponds and lakes.

> *a cloudy morning most of the Chinnooks leave our Camp and return home, great numbers of the dark brant passing to the South* Moulton, *Journals* 6:75 (Clark)

Lewis and Clark were probably a bit surprised and confused by this sudden departure. Later, they would have noticed the high clouds overhead and felt the warm, moist wind. The weather was changing, and it was changing for the worse. The Chinooks recognized the signs of an approaching storm and fled for the shelter of their sturdy homes.

> *The wind blew hard from the S.E. which with the addition of the flood tide raised emence Swells & waves which almost entered our Encampment morng. dark & Disagreeable, a Supriseing Climent*
> Moulton, *Journals* 6:73 (Clark)

The wind rapidly pushed the waters of the Columbia into pointed whitecapped waves. Then, once again, the rain began: not drizzle or thundershowers, but Pacific Northwest rain, large drops falling close together without even the slightest pause. It was the kind of rain that can fill a bucket in just a couple of hours.

Lewis and Clark now recognized that a serious storm was approaching. This was a return to the same kind of weather that had forced them to bury their canoes and sleep on the ground in wet clothes with nothing to eat. It seemed very likely that if they set out now in their heavily loaded canoes, the stormy weather would catch them and put them through a similar ordeal. Obviously, it made more sense to remain right here and wait out the coming storm along this sandy beach than to risk having it hit them along some unknown, inhospitable shore. Wisely, the captains decided to postpone the beginning of their journey home:

Storm clouds build along the coast.

> *at 12 oClock it began to rain, and continued moderately all day, Some wind from the S.E., waves too high for us to proceed on our homeward bound journey* Moulton, *Journals* 6:73 (Clark)

Sgt. Gass's journal echoes this change of plans with some added detail:

> *about 8 o'clock, all the natives left us. The wind blew so violent to day, and the waves ran so high, that we could not set out on our return, which it is our intention to do as soon as the weather and water will permit* Moulton, *Journals* 10:176 (Gass)

Since they were going to remain here, Lewis and Clark found this to be an ideal moment to settle a debt with Sacagawea. She had given them her belt to buy the otter-skin robe, and they wanted to reimburse her for it. The captains opened one of their bundles of baggage, pulled out a coat, and gave it to her.

> *we gave the Squar a Coate of Blue Cloth for the belt of Blue Beeds we gave for the Sea otter Skins purchased of an Indian* Moulton, *Journals* 6:73 (Clark)

This was a generous repayment, and it caught the attention of every member of the party. Clothing was in short supply; most of the men were dressed in rotten, tattered buckskins, so everyone noticed her in the new coat.[1]

Initially, it had appeared as though Sacagawea had lost her belt without any reward, but as it turned out, she ended up with the best part of the exchange. She had traded away an ornamental belt for a valuable cloth coat. During the coming winter she would find the coat to be a great asset, and so would her baby boy, whom she could now wrap up close to her own body.

Despite the worsening weather, a few Indians came to visit Lewis and Clark's camp, giving the captains an opportunity to compile more ethnological data on the Chinooks. Clark wrote extensively about their appearance, clothing, diet, and ornaments. None of these visitors, however, attracted as much attention as one particular businesswoman who arrived at the camp with a troop of young prostitutes.[2]

> *An old woman & wife to a Cheif of the Chinnooks came and made a Camp near ours She brought with her 6 young Squars I believe for the purpose of gratifying the passions of the men of our party*
>
> Moulton, *Journals* 6:75 (Clark)

The Chinooks were traders. They were widely regarded as clever and intelligent businessmen who controlled all transactions in the Lower Columbia. Prostitution was created, no doubt, to satisfy the demand for sex by lonely sailors who arrived in sailing ships. The girls were possibly slaves that had been purchased specifically for this purpose. Clark noticed that one had been tattooed.[3]

Blue Beads

When trading with the coastal Indians Lewis and Clark were somewhat surprised to discover that they valued blue beads more highly than any other object.[1] Lewis experienced this preference when he tried to purchase a sea otter pelt.

> *for these we offered him many articles but he would not dispose of them for any other consideration but blue beads . . . he would not exchange nor would a knife or an equivalent in beads of any other colour answer his purposes, these coarse blue beads are their favorite merchandiz, and are called by them tia Commashuck or Chiefs beads*[2] (Lewis)

These beads were traded from tribe to tribe up the entire watershed of the Columbia River, and were used by the native people as a form of currency.[3]

> *blue beads . . . among all the nations of this country may be justly compared to goald or silver among civilized nations*[4] (Lewis)

I Saw on the left arm of a Squar the following letters J. Bowmon Moulton, *Journals* 6:75 (Clark)

Unfortunately, the sailors had given the girls more than just tattoos. Sailors brought sexually transmitted diseases which soon spread among the Indians. Clark noticed obvious signs of infection which almost certainly had come from contact with foreigners.

maney of the Chinnooks appear to have venerious and pustelus disorders. one woman whome I saw at the beech appeared all over in Scabs and ulsers

Moulton, *Journals* 6:76 (Clark)

Lewis and Clark obviously had every reason to keep their healthy men away from these young girls, but for some reason they didn't. Maybe these particular girls appeared healthy, or maybe their men were so eager that it would have caused a mutiny if the captains had tried to stand in their way. We don't really know. The only concern they mention is their fear that the Chinook girls would persuade the men to trade away valuable steel tools in exchange for sex. To prevent this from happening, the captains gave the men pieces of ribbon from their stock of trade goods.

we divided Some ribin between the men of our party to bestow on their favourite Lasses, this plan to Save the knives & more valueable articles

Moulton, *Journals* 6:74 (Clark)

Clark appears to have been a bit astonished that the Chinooks did not disapprove of sex between these young girls and his men. In fact, he observed the elders' pleasure that these young girls had attracted the men's attention.[4]

The young women Sport openly with our men

Moulton, *Journals* 6:73 (Clark)

The day passed uneventfully. The men amused themselves with the six Chinook girls, but even with this distraction they could never completely forget the worsening weather. Rain continued to pour down harder and harder as dark gray clouds rolled overhead from the south. Waves hammered against the shore. Moment by moment, the storm was growing nearer.

Sick Men

Lewis and Clark should have known that it was unwise to allow their men to have contact with prostitutes and, as one might expect, several of their men become infected with venereal disease. Two months later Lewis writes:

Goodrich has recovered from the Louis veneri which he contracted from an amorous contact with a Chinnook damsel. I cured him . . . by the uce of murcury[1] (Lewis)

(*Louis Veneri* is Lewis's attempt to write syphilis in Latin, which should be spelled *lues venerea.*)

Besides Silas Goodrich it appears that Hugh McNeal also became infected. Lewis writes:

McNeal and Goodrich haveing recovered from the Louis veneri I derected them to desist from takeing the murcury or useing in future. willard is yet complaining and is low Spirited[2] (Clark)

Later, when these girls came to visit Lewis and Clark's men a second time the captains recognized the danger and ordered their men to stay away from them.

this was the Same party which had communicated the venereal to Several of our party in November last, and of which they have finally recovered. I therefore gave the men a particular Charge with respect to them which they promised me to observe[3] (Clark)

Friday, November 22nd

Gale force wind and hard rain pound the coast from daylight until dark. A small high pressure system builds.

Daytime Low Tide:	*7:46 am*	*3.1*
Daytime High Tide:	*1:30 pm*	*8.9*
Sunrise:	*7:23 am*	
Sunset:	*4:41 pm*	

In the early hours of the morning, the storm finally hit. Now, for the very first time, Lewis and Clark's party would feel the full fury of a Pacific Northwest gale.

The winds that accompany these storms blow a steady thirty or forty miles per hour; the gusts hit seventy or eighty. The men's wooden shelters must have creaked and shuddered from the force of each blast.

> *a little before Day light the wind which was from the S S.E. blew with Such violence that we wer almost overwhelmned with water blown from the river* Moulton, *Journals* 6:79 (Clark)

The rain was driven sideways with such force that it blew through the gaps and cracks in the walls of their huts. Within moments, the hurricane-force winds had hurled so much water that everything, whether inside or out, was soaking wet.

> *we are all Confined to our Camp and wet* Moulton, *Journals* 6:79 (Clark)

Hour after hour, the storm raged without a single pause. The exposed beach where the party was camped received the full, uninterrupted force of every gust, some of which may have exceeded ninety miles per hour.[4]

> *this Storm did not Sease at day but blew with nearly equal violence throughout the whole day accompaned with rain* Moulton, *Journals* 6:79 (Clark)

The wind pushed the waters of the Columbia into large waves, which crashed against the shore with tremendous force. Spray flew sixty or eighty feet beyond the water's edge.

Another storm hits the coast.

The waves roared and the wind howled so loudly the men had to shout in each other's ears to make themselves heard.

> *O! how horriable is the day waves brakeing with great violence against the Shore throwing the Water into our Camp* Moulton, *Journals* 6:79 (Clark)

An enormous rogue wave crashed onto the beach, picked up one of their 1,500-pound canoes, and hammered it down with such force that the hull cracked.

> *the waves & brakers flew over our Camp, one Canoe Split by the Tossing of those waves* Moulton, *Journals* 6:79 (Clark)

The energy of these storms is truly terrifying. It humbles even the bravest and boldest individuals. There is nothing one can do except stay low, hang on, and wait.

Saturday, November 23nd

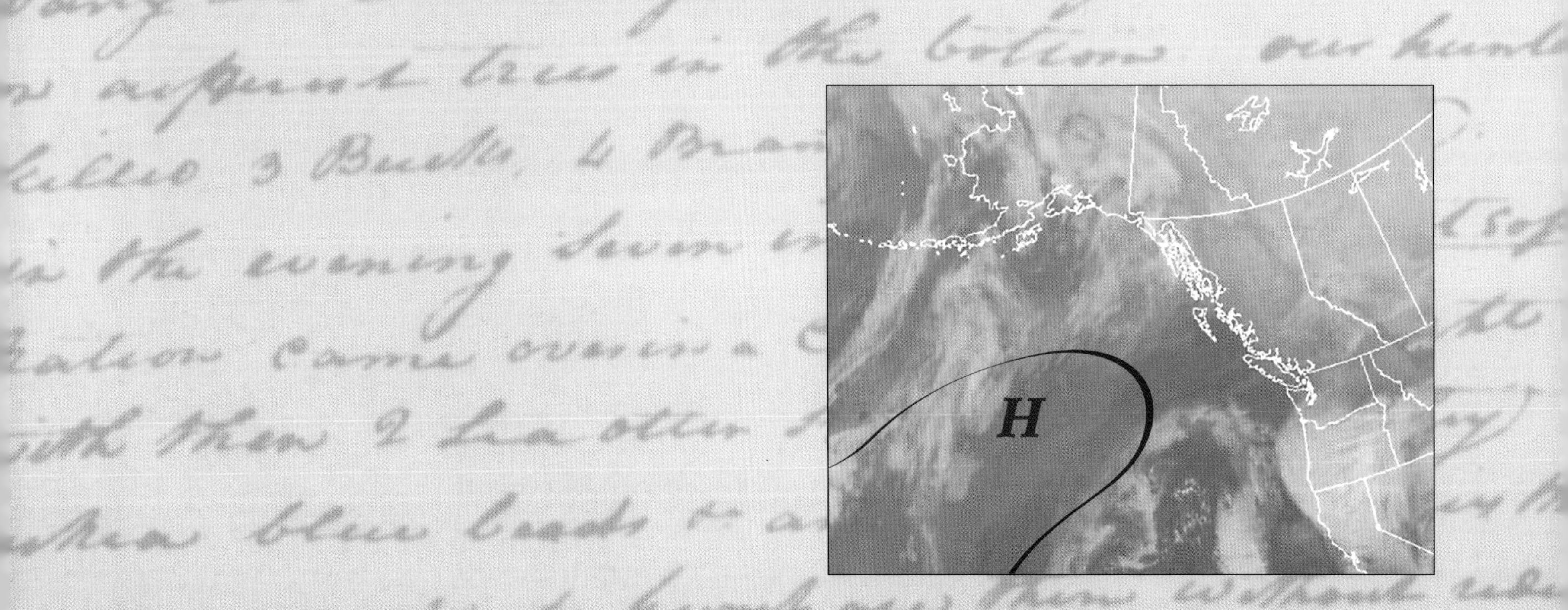

Wind decreases and partial clearing.
High pressure builds.

Daytime Low Tide:	*8:23 am*	*3.3*
Daytime High Tide:	*2:04 pm*	*8.7*
Sunrise:	*7:25 am*	
Sunset:	*4:40 pm*	

The storm diminished during the night. By morning the roaring hurricane-force winds had fallen to a whisper. Huge rolling breakers continued to surge in from the ocean and push the Columbia into steep, pitching waves, but the worst was over.

A calm Cloudy morning Moulton, *Journals* 6:81 (Clark)

The horrific storm had left nothing untouched. The men crawled out from their shelters like weary and battered soldiers who had just gone through a bombardment. It is exhausting to endure these Pacific Northwest winter storms; the noise, wind, rain, and fear drains every ounce of energy.

The men who were skilled at woodworking examined the damaged canoe and immediately began to repair it, while others went hunting for game. The rest of the party had to put the camp back in order. They inspected their baggage, all of which was soaking wet, and spread it all out to dry. Wet sand stuck to everything, and the iron tools were beginning to rust.[5]

While their baggage was being unpacked Captain Lewis noticed his branding iron, and this gave him an idea. He heated it in the fire until the steel was cherry-red hot, then pressed it into the side of a tree.

Capt Lewis Branded a tree with his name Date
Moulton, *Journals* 6:81 (Clark)

Alder trees bordering Station Camp are marked.

"the party all Cut the first letters of their names on different trees in the bottom"
William Clark

Clark didn't have a branding iron, but he did have a knife, so he carved his name into an alder next to Lewis's.

> *I marked my name the Day & year on a Alder tree*
>
> Moulton, *Journals* 6:81 (Clark)

The captains undoubtedly recognized this as a historic site. Others would follow and would come to this same beach where the first American expedition to cross this continent had ended its westward journey. Everyone wanted to be remembered, so, one by one, all the men joined in and each carved his initials into different trees.

> *the party all Cut the first letters of their names on different trees in the bottom* Moulton, *Journals* 6:81 (Clark)

Clatsop Indians from a village near Point Adams paddle across the Columbia bringing furs to trade.

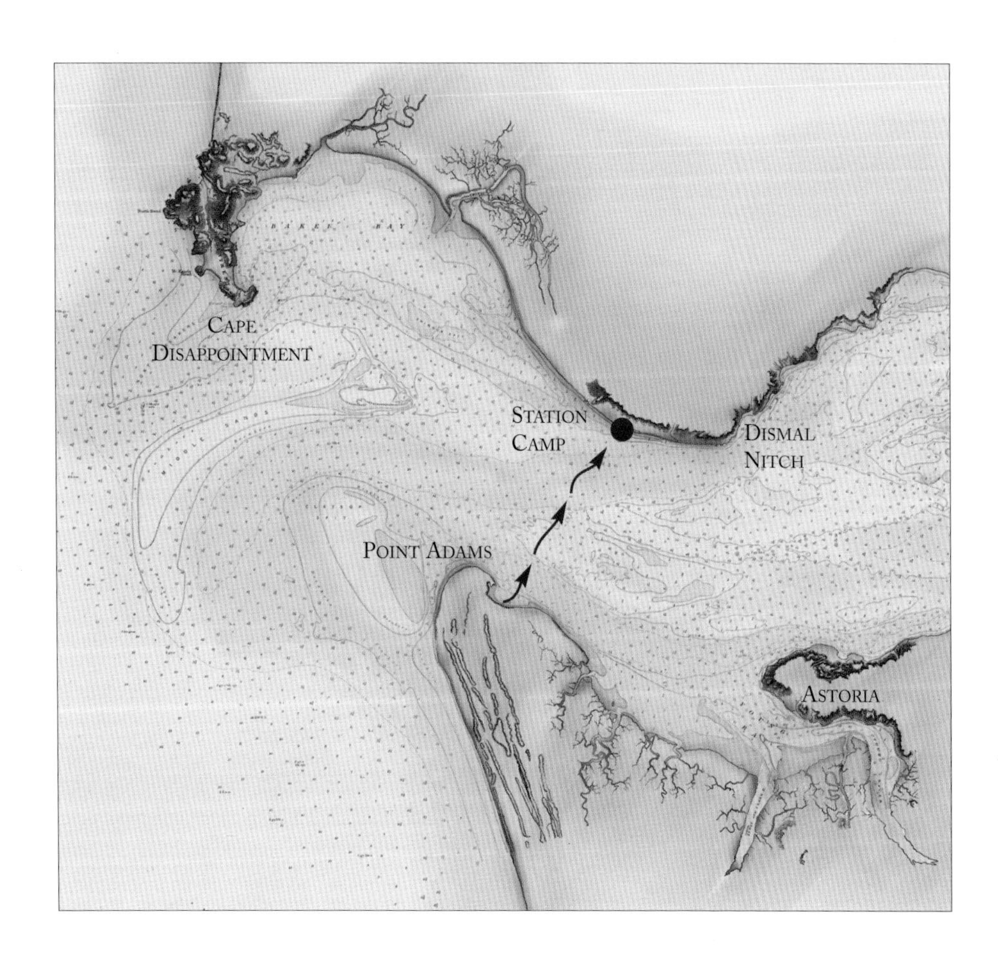

The site was now well marked. Their names would all go down in history.[6]

By late afternoon the broken canoe was repaired and the baggage was repacked. The weather was improving every hour, and, if it continued, Lewis and Clark's party would be able to make their departure the next morning and begin their journey home.

It appeared that this final day near the ocean was going to be uneventful until some Indians paddled across the wide Columbia River and landed beside Lewis and Clark's campsite. These were members of a tribe Lewis and Clark had met only briefly.

> *seven indians of the Clot Sop Nation Came over in a Canoe, they brought with them 2 Sea otter Skins*
>
> Moulton, *Journals* 6:81 (Clark)

It was fortunate that these Clatsops had crossed the river, because Lewis and Clark now had time to record some details about them. They listened closely to the Clatsops when they spoke, they examined their clothing, canoe, and mannerisms, and found, not surprisingly, that this tribe did not much differ from the Indians inhabiting the north side of the river.

> *they Speak the Same language of the Chinnooks and resemble them in every respect*
>
> Moulton, *Journals* 6:81-82 (Clark)

The Indians who had paddled across the Columbia were all Clatsops, but one had different-colored skin and hair. Patrick Gass gives us a brief description in his journal:

> *In the afternoon, 10 of the Clat-sop nation that live on the south side of the river, came over to our camp. . . . One of these men had the reddest hair I ever saw, and a fair skin, much freckled*
>
> Moulton, *Journals* 10:177 (Gass)

The first ship ever known to enter the Columbia River arrived in 1792; however, as this red-headed Indian man plainly proved, contact between the coastal natives and foreigners had occurred long before that. This Clatsop man was approximately twenty-five years of age, but judging by his behavior he had obviously been born among the Indians.

Clark now turned his attention to the sea otter skins the Indians had brought to trade. He wanted to make the purchase, but he had become wary of the coastal Indians. Their appearance was deceptive. Though they wore few clothes and were barefoot, which may have made them seem unsophisticated, they were proving to be extremely shrewd businessmen.

Clark began to suspect that there was no such thing as a set price when dealing with these natives. Their prices rose higher and higher in direct proportion to how much they thought the buyer could pay. Clark began to develop a strategy for bartering with them. He noticed that if a buyer showed excitement over some item, the price invariably was extremely high. If, however, the buyer ignored the same item, he could purchase it for a low price.

To test this theory, since he had already admired these sea otter pelts, Clark laid out a wide array of enticing items in exchange. If his assumptions were correct, the Clatsops would reject everything he included in his first offer. As a test, Clark included his priceless pocket watch in the deal.

> *mearly to try the Indian who had one of those Skins, I offered him my Watch, handkerchief a bunch of red beads and a dollar of the American Coin, all of which he refused*
>
> Moulton, *Journals* 6:81 (Clark)

"great higlers"

Lewis and Clark had no idea that the coastal Indians they were living among were members of an enormous trade network and had been handling merchandise, negotiating, and selling for hundreds, if not thousands, of years.[1] They were expert businessmen.[2]

Governor George Simpson of the Hudson's Bay Company arrived at the mouth of the Columbia in 1824 and witnessed their dealings. Simpson was familiar with many tribes throughout northern North America, and his observations about the Chinooks as traders, following his description of their head-flattening technique, does not leave any question as to what he thought about them:

> *this operation does not seem to give pain as the children rarely cry and it certainly does not affect the brain or understanding as they are without exception the most intelligent Indians and most acute and finished bargain Makers I have fallen in with*[3]

Clark became frustrated when dealing with the Clatsops. He was expecting an established and fair price, and had no idea that the objective of the Indians was to see just how much they could get. He struggled to figure out their pricing system, and in doing so developed a theory:

> *they are great higlers in trade and if they Conceive you anxious to purchase will be a whole day bargaining for a hand full of roots; this I Should have thought proceeded from their want of Knowledge of the Comparitive value of articles of merchindize and the fear of being Cheated, did I not find that they invariably refuse the price first offered them and afterwards very frequently accept a Smaller quantity of the Same article*[4]

All coastal Indians were indeed great hagglers.

Sailing Ships and Disease

Sailing ships have been responsible for spreading diseases around the world since the earliest times. Perhaps the best-known maritime tragedy involved a Genoese trading ship in 1347 that brought bubonic plague to Sicily. The Black Death swept across Europe killing one third of the population, or approximately 20 million people.[1]

The equivalent of the Black Death arrived among Northwest Indians aboard a Spanish sailing ship in the form of smallpox.[2] By 1776, it was ravaging practically every coastal Indian village.[3] The loss of native life is impossible to ascertain; an estimate of 35 percent is considered to be a conservative figure. Some calculate a death rate as high as 80 percent.[4]

A second wave of smallpox arrived in the lower Columbia River around 1800. Lewis noted how its impact devastated the local native people.

> *The small pox has distroyed a great number of the natives in this quarter. it prevailed about 4 years since among the Clatsops and distroy several hundred of them, four of their chiefs fell victyms to it's ravages*[5]

Malaria, measles, influenza, typhus, and typhoid brought more death to the native people.[6] By the 1870s, the northwest Indian population was scarcely 20 percent of its original number. This horrific death of the indigenous people of the Western Hemisphere has been called "one of the greatest demographic disasters in human history."[7]

Sailing ships brought fur traders into remote villages This brought the native people into close contact with any disease the crewmen might have been carrying, which was then passed on to the entire village.

This abandoned Haida house (right) in ruins would have been a common sight following the outbreak of a deadly contagious disease.

The response was exactly what Clark had expected. Even though he had offered more than the pelt was worth, the Indians wanted a little more.

Unfortunately, knowing the secrets of trade among the coastal Indians would not be of much value to Clark because tomorrow they would be leaving these tribes and heading upriver. Their gear was dried, packed, and ready for travel. The men were well rested and fed, the canoe repaired, and the weather calm. Tomorrow they would load their canoes and begin their homeward journey.

The evening became more pleasant than they had ever seen at this camp. There was no wind, and the sun tried to peek between the clouds just as it neared the horizon. The weather was the exact opposite of what it had been just twenty-four hours earlier.

During the night, all the clouds disappeared and the temperatures dropped below freezing. The obsidian-black night sky was filled with millions of stars. The weather had changed once again.

The western sky begins to clear.

Jack Ramsay

This red-headed Indian also attracted the attention of fur traders who lived in the region after Lewis and Clark. Ross Cox, who arrived at Fort Astor in 1812, left us this valuable description of this mysterious coastal Indian.

> *An Indian, belonging to a small tribe on the coast, to the southward of the Clatsops, occasionally visited the fort. . . . and his history was rather curious. His skin was fair, his face partially freckled, and his hair quite red. . . . he was called Jack Ramsay, in consequence of that name having been punctured on his left arm. The Indians allege that his father was an English sailor, who had deserted from a trading vessel, and had lived many years among their tribe, one of whom he married; . . . Old Ramsay had died about twenty years before this period; he had several more children, but Jack was the only red-headed one among them.*[1]

Clark said he was freckled, with long, red hair, and about twenty-five years old. He also observed that Jack Ramsay did not speak English, but appeared to understand more of the words than his fellow tribal members.[2]

CHAPTER SIX

ANOTHER CHANGE OF PLANS

cross & examine

Sunday, November 24th

High pressure moves over the coast. Freezing temperatures in the morning disappear soon after sunrise.

Daytime Low Tide:	*9:01 am*	*3.5*
Daytime High Tide:	*2:39 pm*	*8.4*
Sunrise:	*7:26 am*	
Sunset:	*4:40 pm*	

NOVEMBER 24 WAS A DAY OF BIG, UNEXPECTED changes, and it all began with the weather, which had made a miraculous switch from overcast skies to bright, clear sunshine.

When Lewis and Clark awakened and saw the clear sky, they immediately decided to postpone their departure. Thomas Jefferson wanted to know the exact latitude and longitude of the

During the night the clouds disappeared. The morning was clear, frosty, and bright.

The clear sky provides Lewis and Clark an opportunity to calculate the latitude by use of their sextant.

mouth of every river; however, until now, the stars and planets that they needed in order to make this calculation at Station Camp had been concealed behind the clouds. This sudden appearance of crystal-clear weather gave them a perfect opportunity to use their sextants.

In order to feed the party this one extra day, hunters were sent out in search of game; meanwhile, Lewis and Clark unpacked their navigation equipment and set to work.

> *a fare morning. Sent out 6 hunters and Detained to make the following observations*
>
> Moulton, *Journals* 6:82 (Clark)

Clark used the word "detained" to describe their decision to remain an extra day. Fortunately, Patrick Gass's journal adds clarity to Clark's journal entry:

> *The morning was fine with some white frost. As this was a fine clear day, it was thought proper to*

remain here in order to take some observations, which the bad weather had before rendered impossible. The latitude of this bay was found to be 46° 19 11.7 north Moulton, *Journals* 10.177 (Gass)

Obviously, this decision to delay one extra day near the ocean would not seriously affect the party's survival during the upcoming winter; however, the sudden appearance of frost was worrisome and reminded Clark of their desperate clothing situation.[1] His men were dressed in rags. Finding a winter camp was now a top priority.

being now determined to go into Winter quarters as Soon as possible Moulton, *Journals* 6:85 (Clark)

In choosing a location for their winter camp, Lewis and Clark's main concern was the presence of large herds of deer and elk, which they needed both for food and for hides to make leather clothing.

a convenient Situation to precure the Wild animals of the forest which must be our dependance for Subsisting this Winter Moulton, *Journals* 6:85 (Clark)

So the captains told their men to interview the Indians that milled around the camp to see if they could find out where the best hunting would be found at this time of year.

This Certinly enduces every individual of the party to make diligient enquiries of the nativs the part of the Countrey in which the wild Animals are most plenty Moulton, *Journals* 6:85 (Clark)

The Indians told of one place upriver where the deer were plentiful.

the greatest numbers of Deer is up the river at Some distance above Moulton, *Journals* 6:85 (Clark)

However, the Clatsops who had crossed the evening before had told the men something entirely unexpected. They said that elk were plentiful along the southern shore of the Columbia, directly across from their present camp.

They generaly agree that the most Elk is on the opposit Shore Moulton, *Journals* 6:85 (Clark)

This news would have been very surprising, and some of the men might not have believed it. They had hunted for many days along the north coast and hadn't seen a single elk; now they were being told that lots of elk could be found along the south side of the river. The men could gaze across the broad Columbia and see the area described by the Indians. They must have wondered why the elk gathered

Sextant Navigation

Thomas Jefferson gave specific directions to Meriwether Lewis about calculating, with a sextant, the precise location of key places along their route.

Beginning at the mouth of the Missouri, you will take observations of latitude & longitude, at all remarkeable points on the river, & especially at the mouths of rivers, at rapids, at islands, & other places & objects distinguished by such natural marks & characters of a durable kind[1]

In order to prepare for this navigational task, Lewis studied under the astronomer Andrew Ellicott and the mathematician Robert Patterson. Jefferson recommended a theodolite, but both Patterson and Ellicott disapproved, saying it was too delicate an instrument and liable to become inaccurate. Jefferson conceded and wrote Lewis:

With respect to the Theodolite, I wish you to be governed entirely by the advice of Mr. Patterson & Mr. Ellicott[2]

Lewis and Clark's measurement of 46° 19' and 11.7"[3] is remarkably close to the actual latitude of 46° 14' and 40".[4] An error of only 5' means that they were within five miles of pinpoint accuracy. Considering the fact that the expedition's instruments had been bumped, rolled, tossed aboard boats and horses, heated in the summer sun and frozen in the winter cold, it could be said that Lewis and Clark's precision was nothing less than astounding.

there: did it have a different climate or did different plants grow there?

Until this moment they hadn't given any thought to that side of the river,[2] but it didn't take them long to realize that camping there near the ocean would give them three enormous advantages:

1. The climate was mild;
2. They could boil sea water to make salt;
3. A sailing ship might arrive bringing trade goods they could purchase.

Clark described these advantages as follows:

> *a convenient Situation to the Sea coast where we Could make Salt, and a probibility of vessels Comeing into the mouth of Columbia . . . from whome we might precure a fresh Supply of Indian trinkets . . . added to the above advantagies in being near the Sea Coast one most Strikeing one occurs to me . . . the Climate which must be from every appearance much milder* Moulton, *Journals* 6:85 (Clark)

"greatest number of Deer"

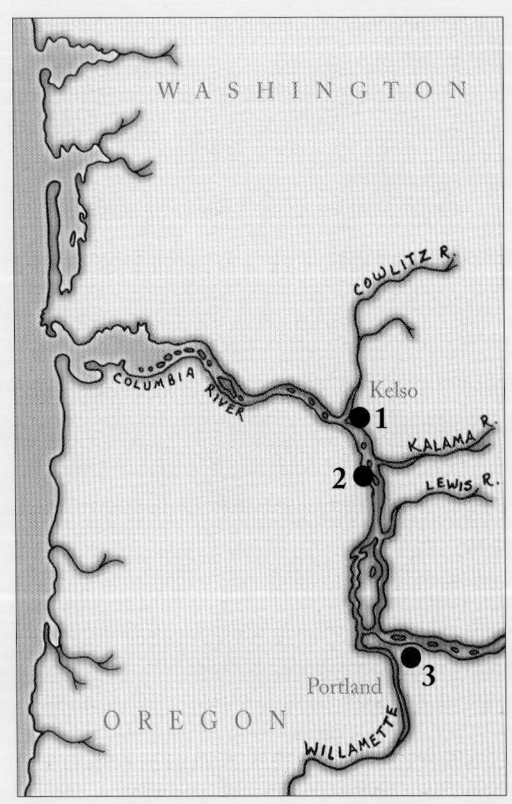

The Indians described good deer hunting upriver, which may have been at one of these sites.

1. *the nativs insisted on our remaining all day with them and hunt the Elk and deer which they informed us was very abundant in this neighbourhood*[1] (Clark)

2. *turned out to hunt very early this morning; by 10 A.M. they all returned to camp having killed seven deer. . . . the hunters informed us that they had seen upwards of a hundred deer this morning on this island*[2] (Clark)

3. *dureing the Short time I remained in their village they brought in three Deer which they had killed with their Bow & arrows*[3] (Lewis)

(opposite) From their camp the men look across the vast Columbia River. Fog hovers low, concealing Youngs Bay. Saddle Mountain rises in the distance.

(left) Black-tailed deer were abundant farther upriver.

The Clatsop Indians tell of elk herds in the vicinity of their village.

Lewis and Clark were obviously attracted to the idea of a winter camp near the ocean, but changing their plans this late in the year could be risky, especially if they based their decision on hearsay. What if they were misunderstanding the Indians? Or worse yet, what if the Clatsops were exaggerating?

The captains were in favor of crossing to search for elk, but they decided to also ask their men what they thought about the idea. So that evening they assembled all the party together and asked to hear each member's opinion. Patrick Gass described the process in these words:

> *At night, the party were consulted by the Commanding Officers, as to the place most proper for winter quarters* Moulton, *Journals* 10:177 (Gass)

This consultation of the men's opinions is one of the most intriguing moments of the entire expedition. Unfortunately, very little information exists concerning this consultation, which has led to a great deal of speculation and a variety of individual interpretations.

What is known about this evening is as follows: Lewis and Clark were trying to decide whether to continue with their plan to build a winter camp somewhere upriver or to change direction and scout the south shore of the Columbia for a winter campsite. In order to find out which one of these two choices was preferred by their party, the captains apparently asked each man to answer a compound question.

The first half of the question focused on the immediate future. It might have been worded, "Where should we go when we leave here tomorrow?"

The second half of the question dealt with a contingency plan. It concerned their second choice of campsites and might have asked, ". . . and where should we go if our first choice turns out to be unsuitable?"

The first man to be questioned was the highest-ranking sergeant, John Ordway. After he responded, the captains asked the other two sergeants. All three men gave identical responses: their first choice was to cross the Columbia and search for elk; if none could be found, their second choice was to go up to the Sandy River.

After each sergeant had spoken, the other members of the party – the privates and interpreters – were asked exactly the same two-part question. Clark wrote down every man's name and how he responded. A majority agreed with the sergeants that the party should cross the Columbia in search of elk.

After the last private had given his response, the captains turned and asked York the same two-part question. What makes this worthy of mention is the fact that York was not formally a member of this party. He was Clark's servant. However, it is obvious that in the course of the expedition York had become much more. He was a skillful, powerful, and trustworthy individual, and if all the facts were known regarding his involvement and contribution we would probably see that he was among the best and most competent of the men. His opinion was considered as valid as any other; Clark recorded his response along with the others.

A clear majority of the members of the party agreed with the plan to cross and search for elk. However, the men were split on the second question. Clark tallied everyone's choice and wrote out the score; six members of the party were in favor of one place, ten men preferred another, and twelve selected yet a different site.

Then, after all of the men had voiced their opinions, a woman's voice was heard. Sacagawea spoke.

How she became involved in this decision is a mystery. It is generally assumed that Lewis or Clark asked to hear her opinion, but we don't know that for a fact. Her comment appears in Clark's journal after everyone else's responses have been tallied. It is entirely possible that the sergeants and privates, upon realizing that she had been left out, asked to hear her opinion. It is equally possible that she just spoke up, without

"my man York"

It was not unusual for military men up through the time of the War Between the States to bring their servants with them on military campaigns; so when Captain Clark brought York with him on this expedition, it probably did not seem unusual to the citizens of the day. However, it is very unlikely that many of the servants taken on these military forays were ever asked for their opinions, and it is even less likely that anyone would have bothered to write their opinions down. The fact that York was consulted and that someone wrote down what he said is what sets the 24th of November apart from any other day in early American slave history.

During the expedition York had become more than just a servant to Clark; in the eyes of the other members of the party he was one of them. York ate the same food, slept on the same ground, suffered the same hardship, and sweated under the same toil. He proved himself reliable time and again.

An example of this can be seen on November 16. After days of hunger Clark sent out a squad of hunters. York went with them. When they returned it was revealed that York had shot twice as much game as anyone else.[1]

Throughout the journals we see this man being sent out to perform tasks along with the other members of the party.

I send out three men to hunt & 2 & my man york in a Canoe ... in Serch of fish and fowl[2]

(Clark)

It is clear that after a year and a half of travel – sharing the grief and joy, the labor and the food – these men were less able to sustain racial prejudices. They were isolated together in the wilderness, trying to make a decision that would guarantee the entire party's survival this winter and insure their safe return in the spring.

being asked. There are a couple of indications in the journals that she was a strong-willed young woman who wouldn't hesitate to confront Clark and tell him exactly what she was thinking.[3]

We know very little about what actually occurred. What we do know for certain is that Sacagawea did express her opinion; she made one clear statement that precisely expressed her preference for a winter campsite. Clark listened, and even though it ran contrary to what every other member of the party had expressed, he wrote down her opinion.

> *in favour of a place where there is plenty of Potas.* Moulton, *Journals* 6:84 (Clark)

Now Captains Lewis and Clark saw that a clear majority of the men agreed with them that their best plan for the immediate future was to cross the Columbia in search of elk. Only John Shields and Sacagawea had different opinions.

> *together with the Solicitations of every individual, except one of our party induced us Conclude to Cross the river and examine the opposit Side*
>
> Moulton, *Journals* 6:85 (Clark)

The decision had been made. Tomorrow they would cross over to the southern side of the Columbia. The river was only five miles wide, which meant they could reach the opposite shore in little more than an hour. If all went well, their hunters would know by tomorrow afternoon whether or not the Indians had given them accurate information.

November 24 had been a very unusual day. The party had awakened that morning thinking that they would load their canoes and begin their journey eastward. However, because of changes in the weather which delayed them one extra day at Station Camp, they had received new information and had made a highly significant last minute change in their plans.

Sacagawea's Response

Sacagawea's response on the night of November 24 is very interesting. What might have been her reasons for saying that they should winter where lots of roots could be found?

We can never know exactly what ran through her mind, of course, but perhaps she thought the men were making a mistake; maybe she thought it was a poor idea to build a camp on the premise that wild animals would be abundant all winter. She may have already experienced a winter where the buffalo had migrated to other areas and left her people hungry. Maybe she knew it was a wiser decision to camp near edible roots where there was a guaranteed meal and where their survival was more likely.

On the other hand, maybe she felt it was her role, as the party's only woman, to contribute to the acquisition of food. She was not an elk hunter and there were very few edible plants available in these coastal woods, so maybe she wanted to camp where roots grew abundantly and could be gathered as her contribution to the party's diet.

Or perhaps she just wanted to be in the best position to insure the survival of her baby. After all, if something happened to all these men, what would her options be? What would she eat? She wasn't a hunter, but she would have realized that if they were living near an abundant supply of edible plants she could always find something to make a meal.

It could also have been simply a matter of preference for the taste of wappatoe roots. (Clark tells us that in his opinion it was a good substitute for bread.)[1]

The real significance of her opinion may be that she was such an independent thinker that she could not be persuaded by the overwhelming majority of the men. Twenty-seven men had already voiced their intention to head south in search of elk, but Sacagawea spoke her mind and said she felt they should go back upriver to the land of the wappatoe.

The Vote

To reach their decision about the vital matter of where to establish a winter camp, Lewis and Clark took a vote which is considered to demonstrate the first real enfranchising of a black man and an Indian woman in the country's history.

On the evening of November 24, Lewis and Clark found themselves in an odd predicament. New information had opened the possibility of remaining near the ocean for the entire winter, but this created a dilemma. Should they change their plans and cross the Columbia, or should they return upriver to where they had previously decided to go?

To help resolve this issue, Lewis and Clark gathered the entire party together to "consult the men's opinions." Their experiences had not, of course, been identical; some had been out hunting whereas others had spent more time in camp with visiting Indians. Gathering the entire party together and giving everyone a chance to speak would be the best procedure for weighing each person's information.

When this consultation of opinions was over, Lewis and Clark realized that an overwhelming majority of the party felt as they did. The decision was now easy. Tomorrow they would pack up, cross the river, and search for elk.

"Consultation of the Men's Opinions"

November 24 was not the first time nor the last time Lewis and Clark would consult their men's opinions. The first recorded occasion occurred on August 4, 1804, when the captains asked the men to vote on the person they wanted to be promoted to the rank of Sergeant after the death of Sergeant Floyd.

> *ordered a vote for a Serjeant to chuse one of three which may be the highest number the highest numbers are P. Gass had 19 Votes, Bratten & Gibson*[1] (Clark)

The next noteworthy occasion took place when Lewis and Clark's party arrived at the junction of the Maria and Missouri Rivers. No one was sure which river was the Missouri. Several days were spent trying to decide, and during this time there appears to have been some sort of informal polling of everyone's opinion.

> *the party with very few exceptions have already pronounced the N. fork to be the Missouri; myself and Capt. C. . . . have not yet decided but if we were to give our opinions I believe we should be in the minority.*[2] (Lewis)

During their return trip, one of the party asked to leave the expedition and join up with some trappers. Lewis and Clark agreed to this on the condition that everyone else would remain with them to St. Louis. This was discussed and agreed upon by the men.

> *Colter one of our men expressed a desire to join Some trappers . . . we agreed to allow him the prvilage provided no one of the party would ask or expect a Similar permission to which they all agreed*[3] (Clark)

Explorers often relied on scouting parties to gather information that would be used to make decisions, and Lewis and Clark were no different. They were surrounded by alert, intelligent, and perceptive men whose opinions they trusted and often relied upon. However, it was not common to gather everyone together in one place at one time to ask their personal views; and what sets November 24 apart from all the other consultations of the men's opinions was the fact that Clark listed everyone's name and carefully recorded each person's response.

The men's opinions as they appear in Clark's journal on November 24, 1805

Sergt J. Ordway [1]	Cross & examine [2]		S [3]
Serjt. N. Pryor [4]	do [5]	do	S
Sgt. P. Gass [6]	do	do	S
Jo. Shields [7]	proceed to Sandy R.		
Go. Shannon	Examn.	Cross	falls [8]
T.P. Howard	do	do	falls
P. Wiser	do	do	S. R. [9]
J. Collins	do	do	S. R.
Jo Field	do	do	up [10]
Al. Willard	do	do	up
R Willard [11]	do	do	up
J. Potts	do	do	falls
R. Frasure	do	do	up
Wm. Bratten	do	do	up
R. Field [12]	do	do	falls
J. B: Thompson	do	do	up
J. Colter	do	do	up
H. Hall	do	do	S. R.
Labeech	do	do	S R
Peter Crusatte	do	do	S R
J. B. Depageu [13]	do	do	up
Shabono [14]	—	—	—
S. Guterich	do	do	falls
W. Werner	do	do	up
Go: Gibson	do	do	up
Jos. Whitehouse	do	do	up
Geo Drewyer [15]	Examn other side		falls
McNeal	do	do	up
York [16]	"	"	lookout [17]

falls	Sandy River	lookout up [18]
6	10	12

Janey [19] in favour of a place where there is plenty of Potas.[20]

Moulton, *Journals* 6:83-84 (Clark)

1. The first to be consulted is the party's senior sergeant. He is the highest-ranking man below Lewis and Clark.
2. Ordway wants to paddle across to the other side of the Columbia and search to see if elk can be found there.
3. *S* refers to Ordway's second choice, which is short for the *Sandy River.*
4. The most senior man beneath Ordway is consulted next.
5. This is Clark's shorthand for *ditto*.
6. The third sergeant responds in perfect unison with the first two. It is interesting to note that Clark abbreviates *sergeant* three different ways.
7. John Shields completely ignores the sergeants and chooses to go directly upriver 120 miles to the Sandy River.
8. *Falls* refers to the Celilo Falls, two hundred miles upriver.
9. *S. R.* refers to the Sandy River.
10. *Up* probably refers to the vicinity of The Dalles, or somewhere downriver from there.
11. Clark misspells the name of *Richard Windsor.*
12. Reuben Field votes differently from his brother Joseph.
13. Clark confuses the name of *Lepage*.
14. Charbonneau remains silent.
15. George Drouillard's response is the same as the men before him; however, Clark does not use *ditto* when he records how he replies. Drouillard is unquestionably the most valuable member of the party, and it appears that Clark made a point of writing down his suggestion.
16. York is Clark's personal servant. Technically he is a slave, but here Clark consults his opinion.
17. *Lookout* might refer to their *Rock Fort Camp* at The Dalles.
18. Clark tallies up the three most popular second choices, but he makes an error in arithmetic. He accurately adds up six for the *Falls* but he gives the *Sandy River* one extra point. And he shorts the *lookout up* column by one. The real result of this consultation is:

Fall	6
Sandy River	9
lookout up	13

19. Sacagawea's name was sometimes spelt:
    Sah-ca-gar me ah
    Sar car gah we a
    Sah-cah-gar-we-ah
    Sah-cah-gah, we a
    Lewis and Clark also refer to her as: *the interpreters wife, the Squar, the Indian woman*, and *the Squar wife of Shabono.* However, on two occasions Clark refers to her as *Janey* which merely adds to the confusion over the pronunciation of her name.
20. *Potas* undoubtably refers to the wild root called wappatoe. Clark mentions that when roasted it tasted like *Irish Potatoes.*

CHAPTER SEVEN

In Search of Elk

our wish which is to examine if game Can be procured

Monday, November 25th

Coastal winds increase as another front moves across the ocean. Overcast but dry.

Daytime Low Tide:	*9:41 am*	*3.6*
Daytime High Tide:	*3:16 pm*	*8.1*
Sunrise:	*7:27 am*	
Sunset:	*4:38 pm*	

Lewis and Clark's decision to "cross & examine" the south shore of the Columbia River had seemed like a good idea the night before. However, when the men awakened and looked out at the vast Columbia, they realized that crossing miles and miles of open water in their heavy canoes with the surge from the ocean hitting them broadside would be risky.

> *The Wind being high rendered it impossible for us to Cross the river from our Camp*
>
> Moulton, *Journals* 6:87 (Clark)

Everyone knew that farther upriver the Columbia was less than a mile wide. Though it might take several hours to get there, it would certainly be a safer place to cross.

> *we deturmind to proceed on up where it was nar-row, we Set out early accompanied by 7 Clat Sop*
>
> Moulton, *Journals* 6:87 (Clark)

The Clatsop Indians led the party upriver. Their canoes glided along the shore, around Point Distress, and passed Dismal Nitch where they had been stuck during those dreadful days from November 10 to 14. They were heading east for the first time and passing places they had already seen, which was an entirely new experience for the men.

Twelve miles ahead was the first narrow point along the Columbia. The Clatsop knew that even children could cross the river here, so they turned and led the way across to the south shore.

> *they left us and Crossed the river through emence high waves*
>
> Moulton, *Journals* 6:87 (Clark)

Lewis and Clark wanted to follow them, but the waves were too dangerously high for their heavy, squat canoes.

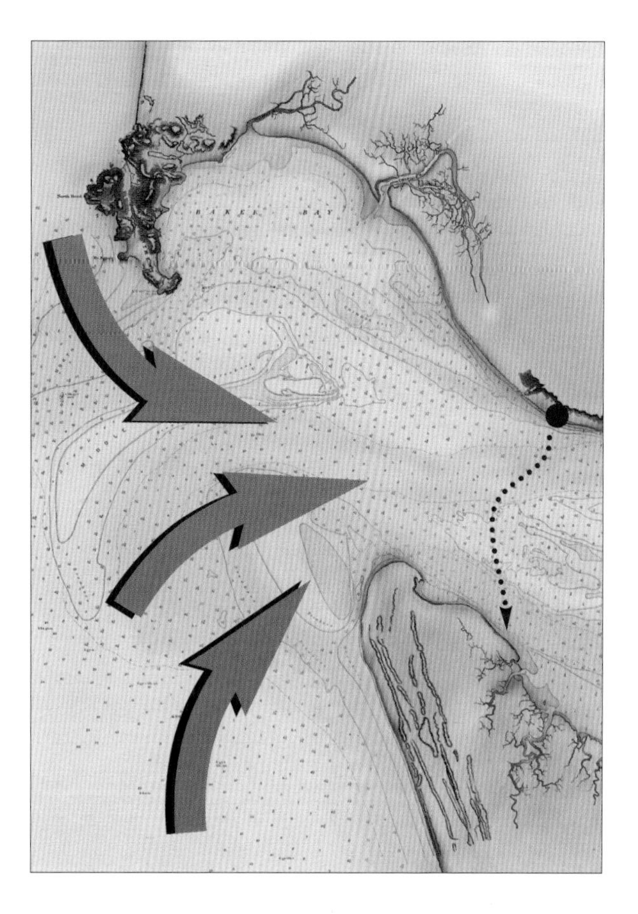

The waves at the mouth of the Columbia prevent Lewis and Clark from crossing directly to where the Indians said elk were most abundant.

(below) Led by the Clatsop Indians, Lewis and Clark's party leave Station Camp and return upriver along the shore.

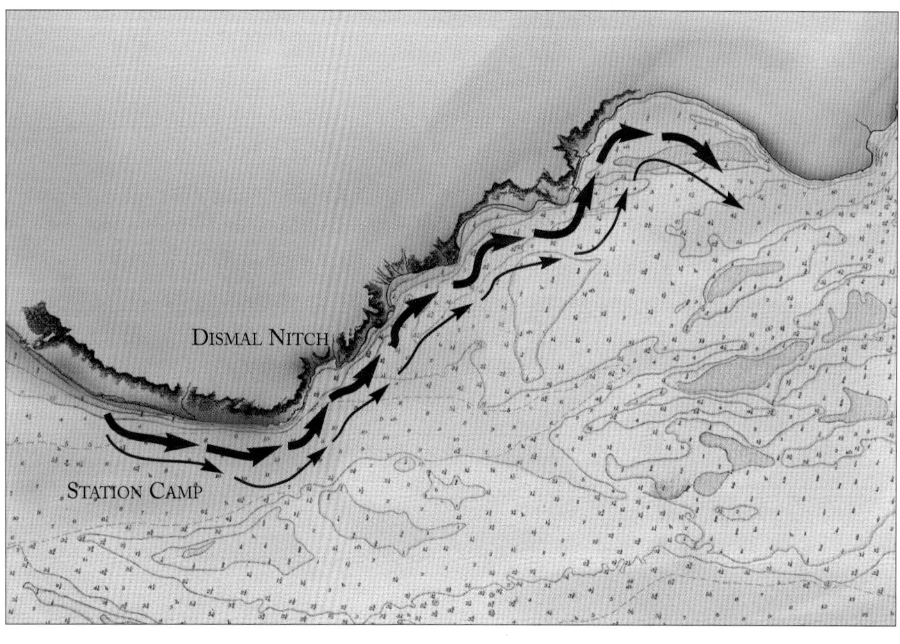

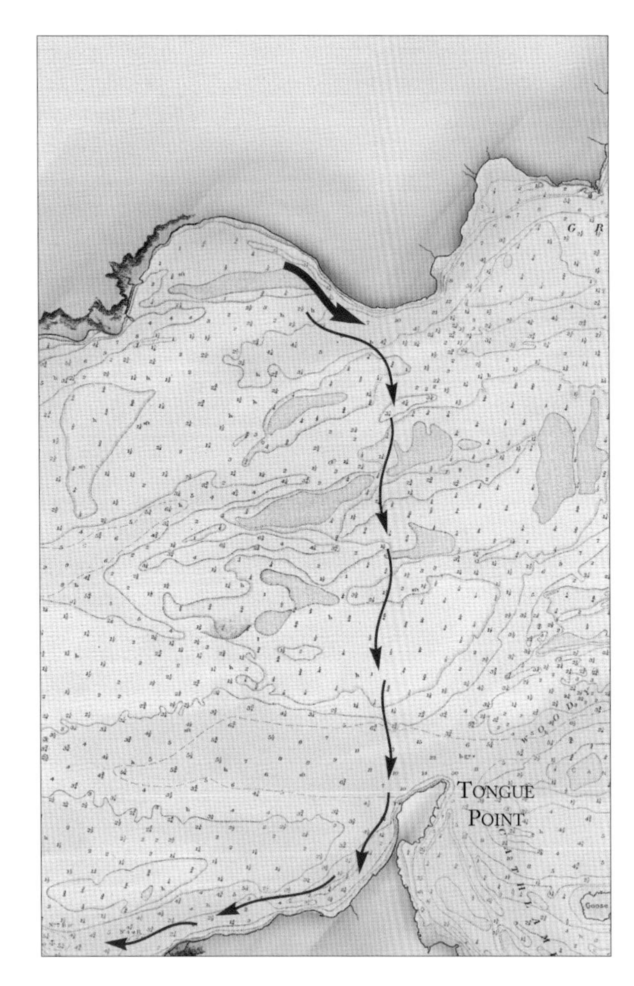

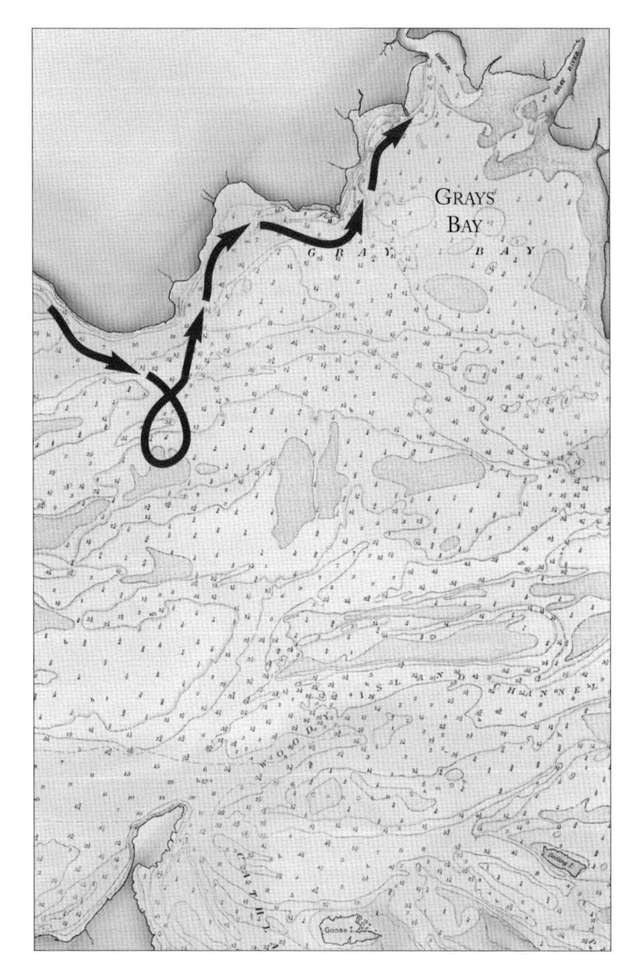

(left) The Clatsops cross the river at the first narrow point.

(right) Lewis and Clark find the waves too high, so they continue upriver hugging the shore.

Looking east. The party paddles upriver in search of a narrow point to cross.

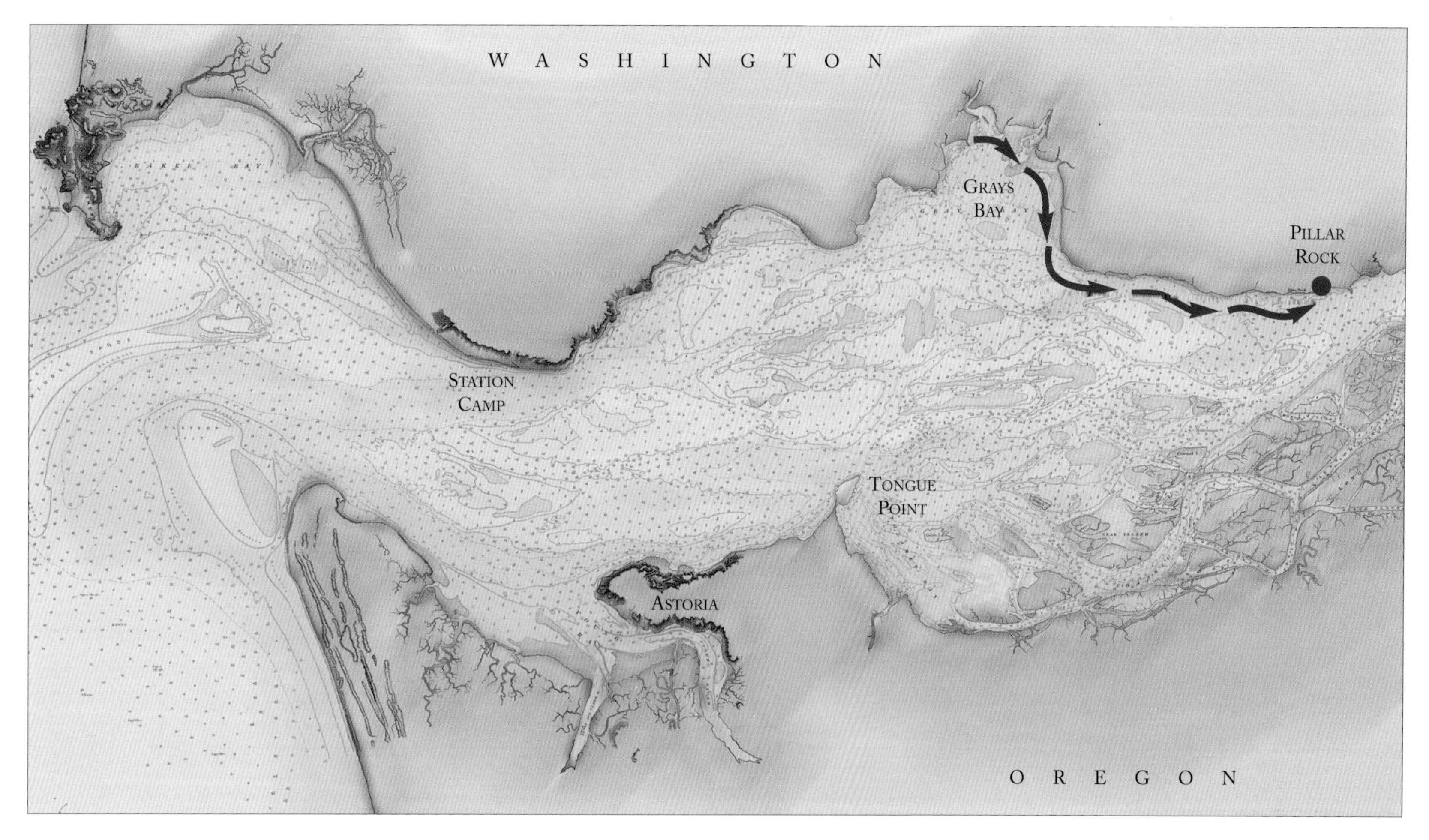

The Swells too high to cross the river, agreeabley to our wish which is to examine if game Can be precured Sufficent for us to winter on that Side

Moulton, *Journals* 6:86-87 (Clark)

After stopping to eat their midday meal,[1] the party continued on, paddling around the shore of Grays Bay and upriver along the shore. As evening approached, they found themselves in nearly the same campsite where they had been nineteen days earlier. This one day of travel had gone remarkably well.

we proceeded on near the North Side of the Columbia, and encamp a little after night near our Encampment of the 7th Moulton, *Journals* 6:87 (Clark)

They had covered twenty miles and were in a perfect position to cross the Columbia, which was still wide but was filled with numerous islands. If the weather remained calm, they could slip across early tomorrow.

Once again the men could gaze downriver and out into the ocean, exactly as they had done on November 7. John Shields and Sacagawea wanted to continue on upriver, but the decision had been made. Tomorrow they would return back into the mouth of this great river.

The party continues upriver all afternoon. They beach their canoes at practically the same place they camped on the night of November 7. The river here is narrow.

Tuesday, November 26th

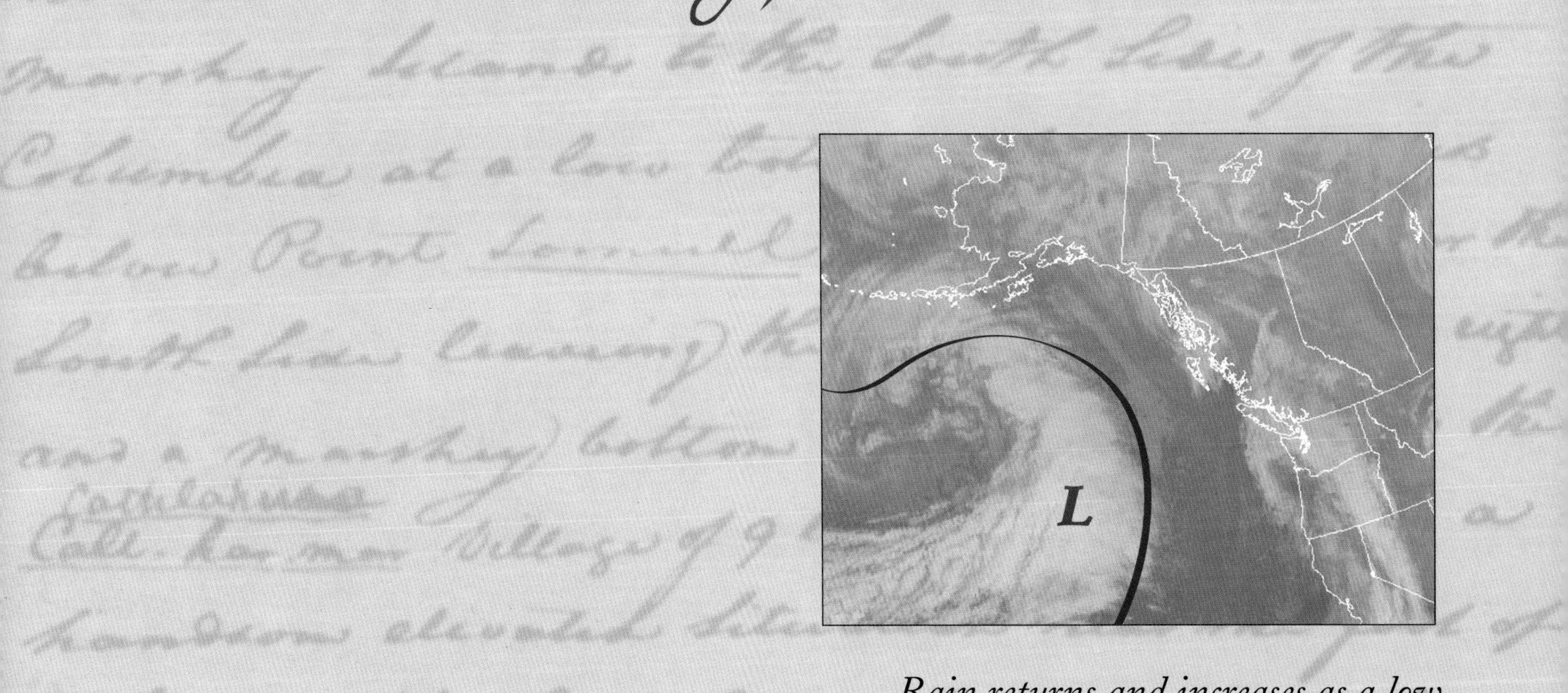

Rain returns and increases as a low pressure system approaches the coast.

Daytime Low Tide:	*10:26 am*	*3.6*
Daytime High Tide:	*3:58 pm*	*7.8*
Sunrise:	*7:29 am*	
Sunset:	*4:37 pm*	

Lewis and Clark took advantage of the calm weather that generally occurs in the early hours of morning. They loaded their canoes, paddled a short distance, turned and glided across to an island without any difficulty.

> *we Set out and proceeded on up on the North Side of this great river to a rock in the river from thence we Crossed to the lower point of an Island*
>
> Moulton, *Journals* 6:87 (Clark)

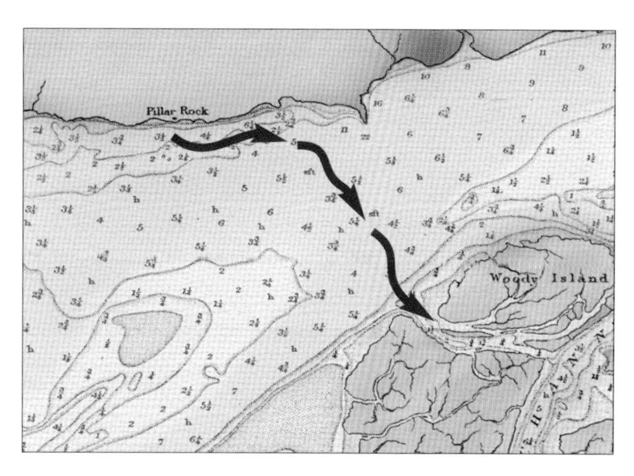

The party paddles farther upriver, then crosses to an island.

They skirted this island, passed between two others, and within minutes were safe along the southern shore. This crossing of the Columbia went like clockwork.

> *passed between 2 Islands to the main Shore, and proceeded down the South Side*
>
> Moulton, *Journals* 6:87 (Clark)

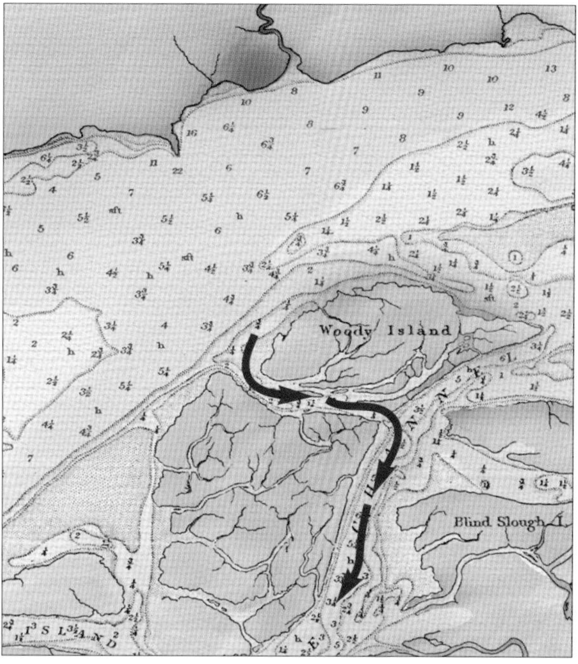

Their exact route is unknown; they would have selected the most direct course that took them downriver.

The calm, smooth waters allowed all five canoes to move quickly along without delay. After several miles they arrived at an Indian village built on top of a ridge. The natives appeared friendly, so Lewis and Clark agreed to land there and eat breakfast.

> *to the Calt-har-mar Village of 9 large wood houses on a handsom elivated Situation near the foot of a Spur of the high land. . . . We purchased*

(bottom) The view looking west straight down the Columbia. The Pacific Ocean is seen in the far distance. Lewis and Clark paddle among the islands in the foreground.

Near the mouth of a river they find a large village. The inhabitants are willing to sell some food, so Lewis and Clark purchase a meal for their men.

(opposite) In the afternoon they continue downriver, following along the quiet waters that lie between the shore and numerous islands.

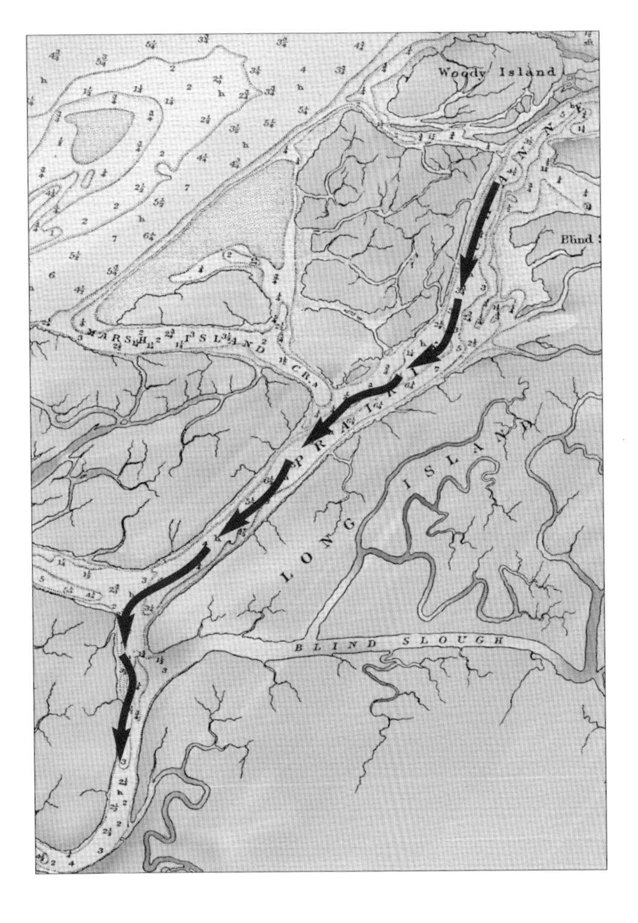

> *Some green fish, & wap pa to for which we gave Imoderate pricie's* Moulton, *Journals* 6:89 (Clark)

Once again Lewis and Clark took on the role of ethnologists. They listened closely to how these Indians spoke and examined their clothing, customs, diet, and houses. As one might expect, these Indians were remarkably similar in every way to the other tribes on the Lower Columbia.

> *This nation appear to differ verry little either in language, Customs dress or appearance from the Chin nooks & War-ci a cum*
>
> Moulton, *Journals* 6:89 (Clark)

After the men were fed and rested, they loaded their five canoes and continued downriver, paddling along between the islands and the shore. They passed one curious island that served as a cemetery. Clark observed the sacred site from a distance and noted the burial canoes elevated on posts above the ground.

(below) The party now looks across the wide Columbia towards the northern shore.

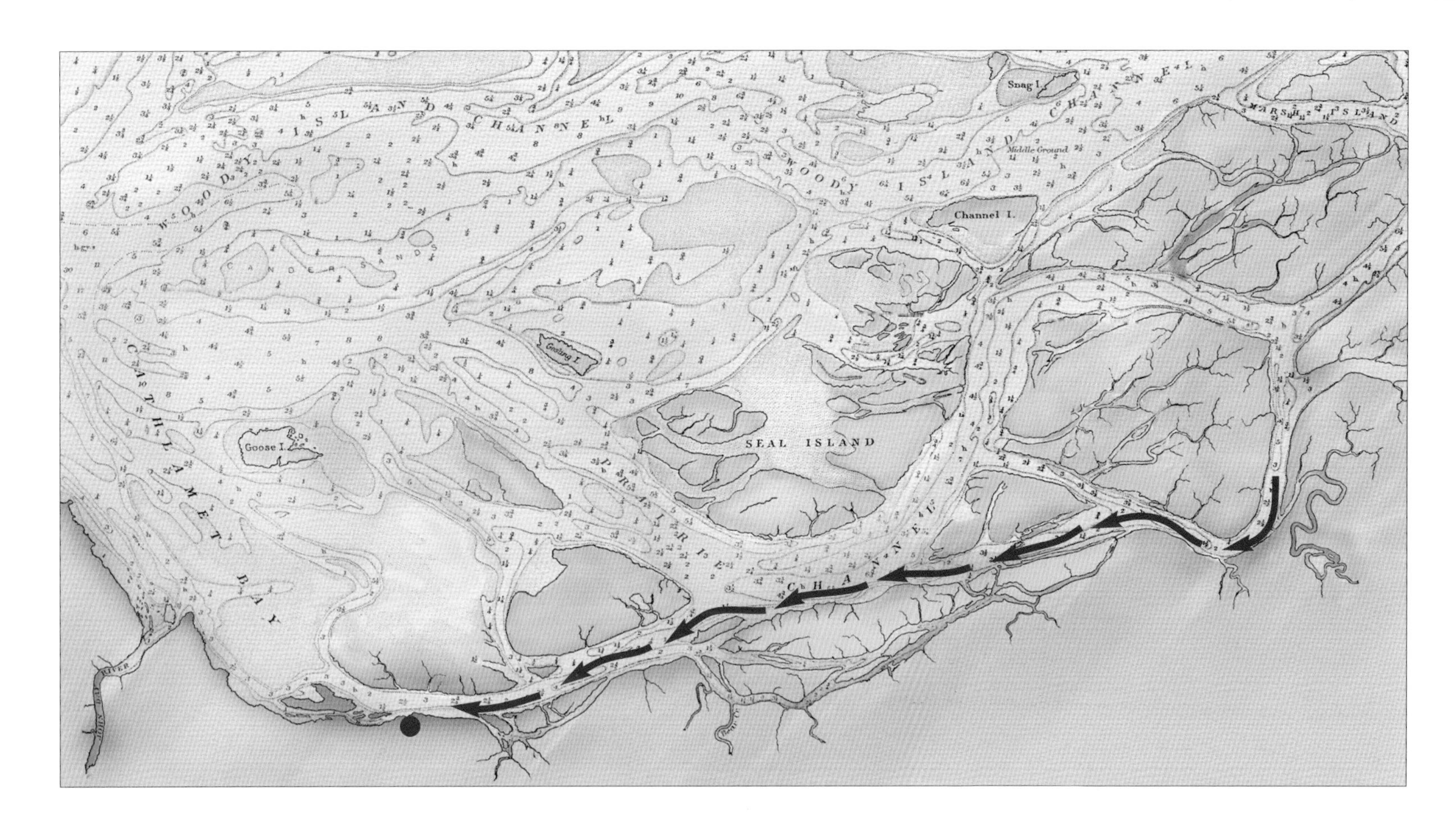
Snag I.
Middle Ground
Channel I.
SEAL ISLAND
Goose I.

a Short distance below the Calt har mer Village on the Island which is Opposit I observed Several Canoes Scaffold in which Contained their dead

Moulton, *Journals* 6:89 (Clark)

As they proceeded downriver, Lewis and Clark saw that the Columbia's shoreline alternated between steep, rugged hills and swampy marshes. It was an inhospitable place, but nevertheless they were forced to find a campsite. Finally, after considerable searching, they pulled ashore along the narrow bank. Rain had been falling all day. Everyone was soaking wet and miserable.

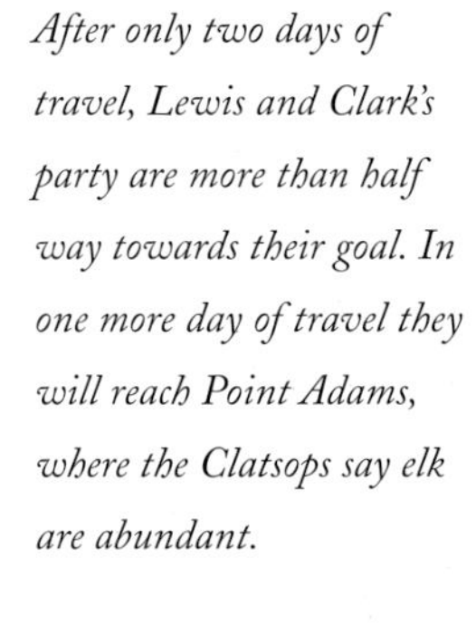

After only two days of travel, Lewis and Clark's party are more than half way towards their goal. In one more day of travel they will reach Point Adams, where the Clatsops say elk are abundant.

We proceeded on about 8 miles and Encamped in a deep bend to the South, . . . all wet and disagreeable a bad place to Camp all around this great bend is high land thickly timbered brushey & almost impossible to penetrate

Moulton, *Journals* 6:88 (Clark)

Lewis and Clark now realized that there was a difference between the northern side of the Columbia and this southern side. The north shore occasionally received a little sunlight, which helped to dry out the terrain; the southern shore was in perpetual shade, and thus remained cold, dark, and soaking wet.

we found much difficuelty in precureing wood to burn, as it was raining hard, as it had been the greater part of the day

Moulton, *Journals* 6:89 (Clark)

As darkness approached, Lewis and Clark's party once again wrapped up in their blankets and settled in for the night. Despite being cold and wet, their spirits must have been high. In two days, they had traveled nearly fifty miles. At this rate, by tomorrow they would be feasting on fat elk for dinner – that is, if they could find any of the elk the Clatsops had described.

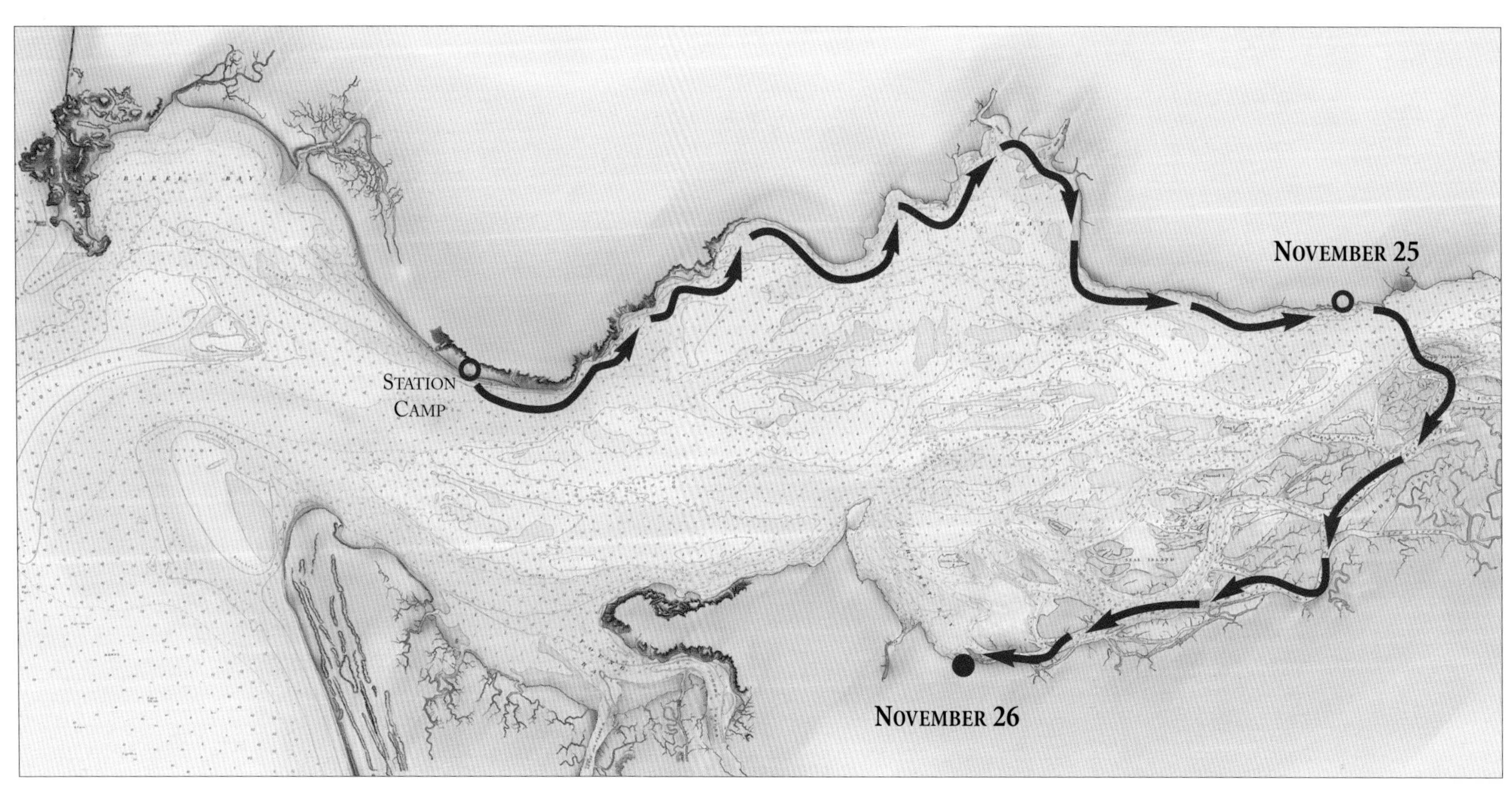

Indian Burial

The Indians of the lower Columbia did not inter their deceased family members. Instead, they constructed platforms elevated above the ground. Lewis described this burial technique.

> *The Clatsops Chinnooks &c. bury their dead in their canoes. for this purpose four pieces of split timber are set erect on end, and sunk a few feet in the grown . . . through each of these perpendicular posts, at the hight of six feet . . . two bars of wood are incerted; on these cross bars a small canoe is placed in which the body is laid after being carefully roled in a robe of some dressed skins; a paddle is also deposited with them; a larger canoe is now reversed, overlaying and imbracing the small one, and resting with it's gunwals on the cross bars*[1]

Lewis tried to understand their beliefs but found that this was impossible without a more fluent understanding of their language.

> *I cannot understand them sufficiently to make any enquiries relitive to their religeous opinions, but presume from their depositing various articles with their dead, that they believe in a state of future existence*[2]

Wednesday, November 27th

Rain and gusty winds continue, hard at times.

Daytime Low Tide:	*11:19 am*	*3.6*
Daytime High Tide:	*4:47 pm*	*7.3*
Sunrise:	*7:30 am*	
Sunset:	*4:36 pm*	

Lewis and Clark's party suffered through another long, wet, miserable night, but the prospect of a nearby winter camp must have boosted their spirits. In another dozen miles they would be walking the ground where the Clatsops had assured them elk were abundant.

> *Rained all the last night and this morning it Continues moderately* Moulton, *Journals* 6:90 (Clark)

Before they could load up and set out, some Cathlamet Indians arrived in camp with a load of merchandise to sell.

> *at day light 3 Canoes and 11 Indians Came from the Village with roots mats, Skins &c. to Sell, they asked Such high prices that we were unable to purchase any thing* Moulton, *Journals* 6:90 (Clark)

During the negotiation, one of the Indian merchants picked up a camp axe without being seen and hid it under his clothes. However, Lewis and Clark's men were so astute that, before setting out, they checked their inventory of supplies and noticed they were one axe short.

> *as we were about to Set out missed one of our axes which was found under an Indians roab*
>
> Moulton, *Journals* 6:90 (Clark)

The party did not have an abundance of tools and could not afford to lose a single item. They closely watched every knife, kettle, and sharpening file. Clark scolded the thief and sent all the Indians back to their village.

> *I smamed [shamed] this fellow verry much and told them they should not proceed with us*
>
> Moulton, *Journals* 6:90-91 (Clark)

Then, without further delay, the men loaded up and proceeded downriver. They

They continue along the shore, pass the mouth of a river and reach a point of land that curves out into the river.

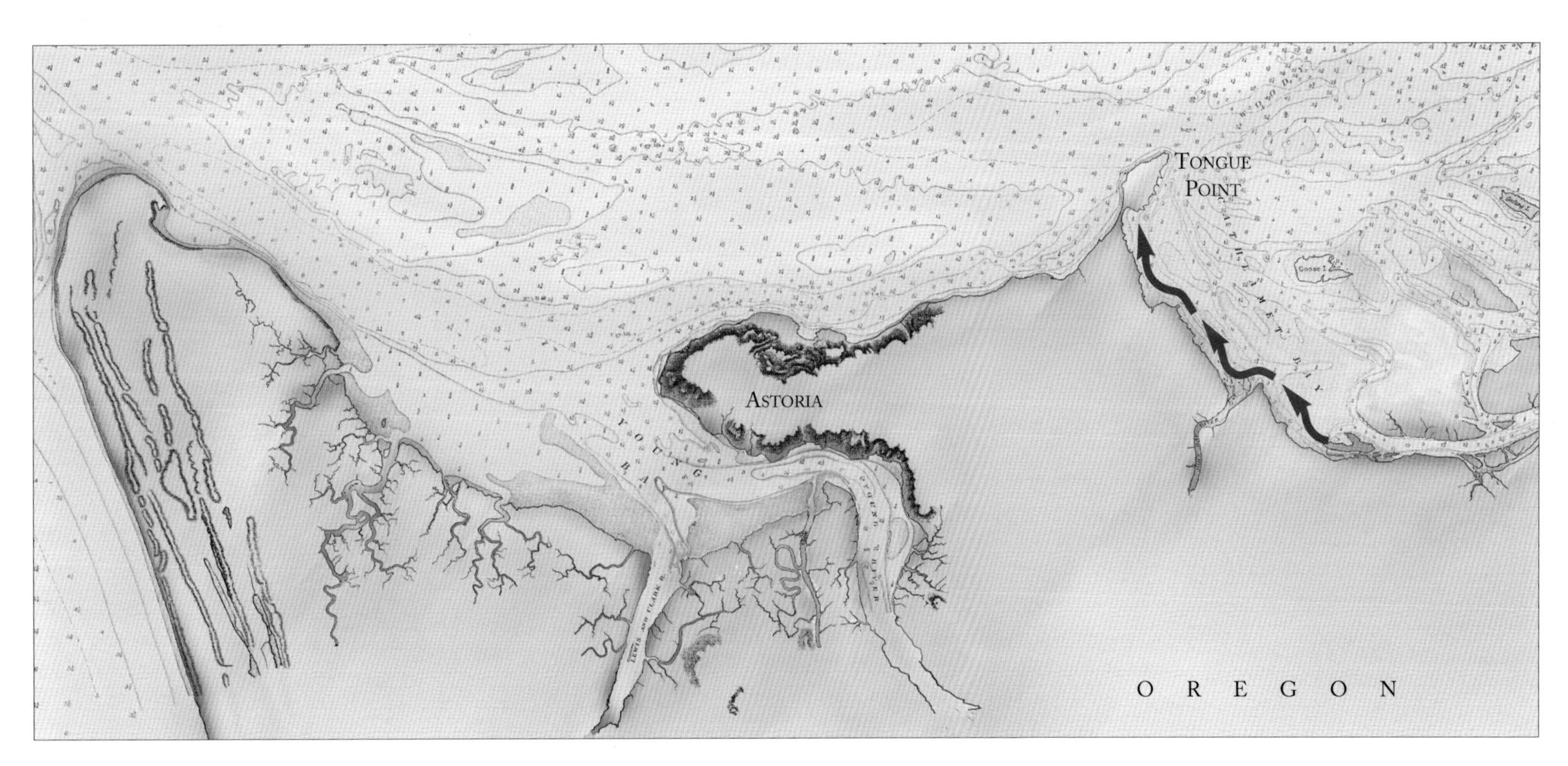

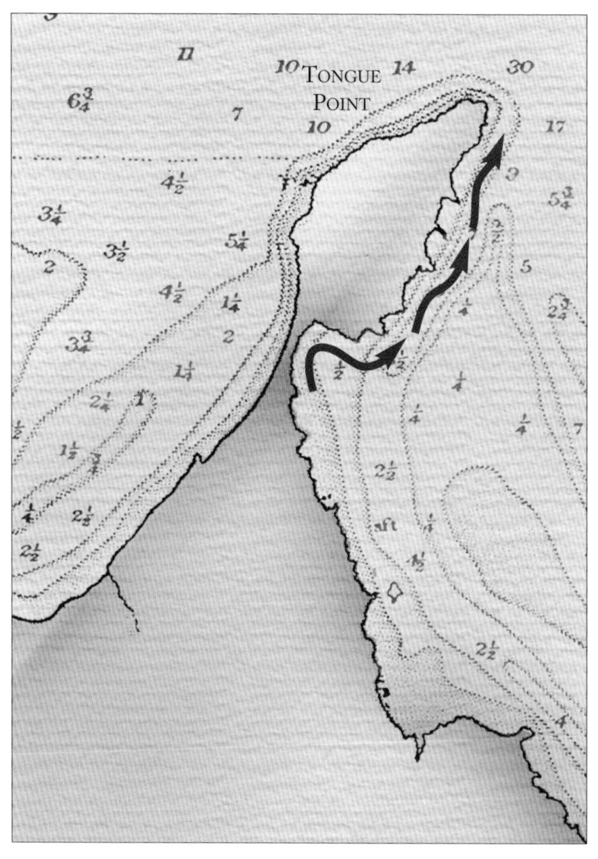

passed the mouth of a stream, skirted several low, swampy islands, and arrived at an abrupt curve in the shore.

> *we proceeded on between maney Small Islands passing a Small river* Moulton, *Journals* 6:91 (Clark)

Here a ridge of rock, covered with thick forest, thrust directly out into the Columbia, creating a teardrop-shaped peninsula. Lewis and Clark led their fleet of canoes out and around this curious point of land.

> *and around a verry remarkable point which projects about 1 ½ Miles directly towards the Shallow bay* Moulton, *Journals* 6:91 (Clark)

Their day of travel went exactly as planned, until they rounded the tip of the point. Here

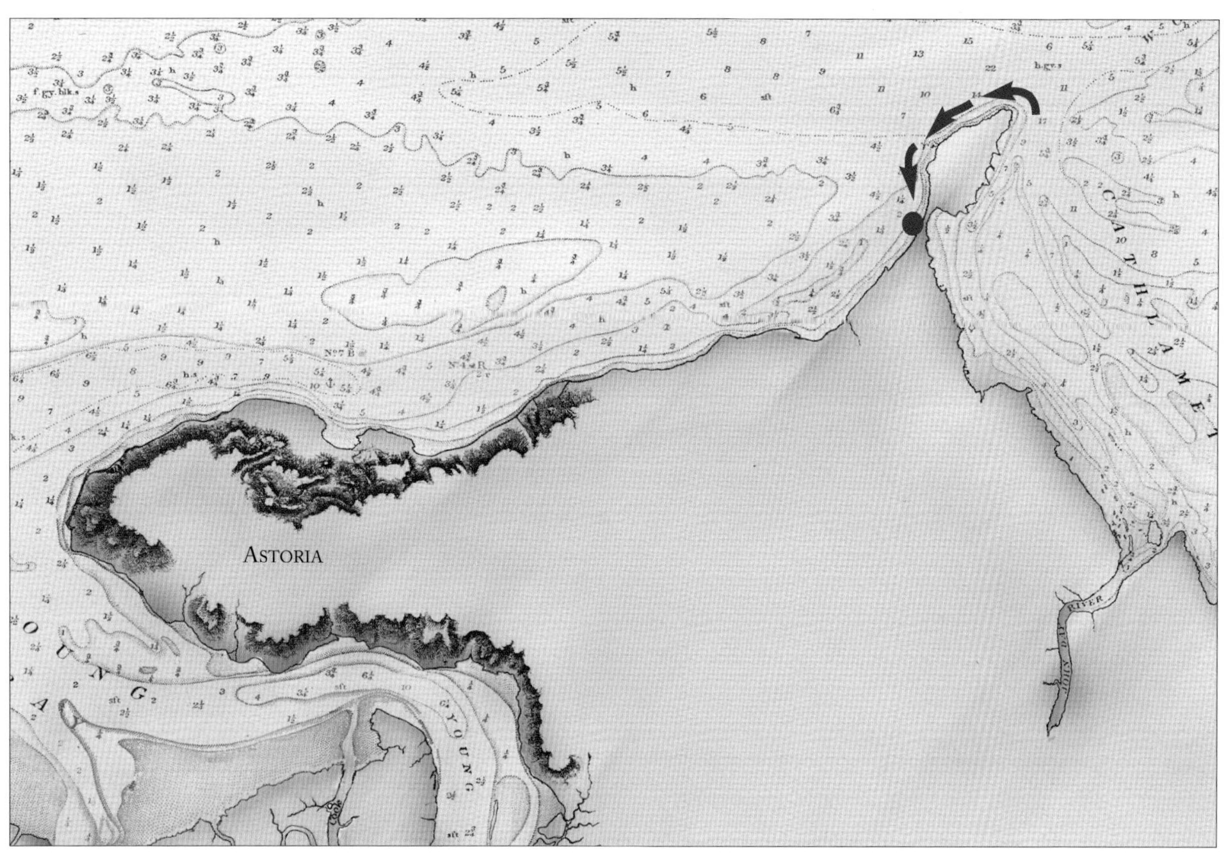

(opposite, left) High waves break around Tongue Point.

(opposite, right) The calm, protected waters are ideal for canoe travel; the party rapidly advances to the tip of the point.

(left) Around the point, the ocean surge hits the tiny fleet, threatening its destruction. The men land at the nearest beach, abruptly ending their travels for the day.

they met the one thing they had hoped to avoid: they ran head-on into the all-too-familiar surge of ocean swells.

The high, pitching waves tossed and rolled their fleet of canoes like leaves in the wind. The crews fought hard to keep their canoes moving downriver, but, in the end, it was just too dangerous. Cautiously, the captains led the party ashore.

> *below this point the waves became So high we were Compelled to land* Moulton, *Journals* 6:91 (Clark)

The wind increased and so did the waves, which hammered their canoes against the shore. They quickly unloaded their gear but, despite their efforts, the waves slammed one canoe into the rocks with such force the hull split.

> *Soon after our landing the wind rose from the East and blew hard accompanied with rain, this rain obliged us to unload & draw up our Canoes, one of which was Split to feet before we got her out of the river* Moulton, *Journals* 6:90 (Clark)

Once again they had been caught by the ferocious, unpredictable Columbia River. Everyone must have been stumbling around in utter disbelief. They had been expecting a fine supper of roasted elk this evening, but instead ended up three miles short of their destination and immobilized. The rain poured down by the bucketful. The men wrapped up in their wet blankets and shivered through another night.

Thursday, November 28th

Gale force winds and hard rain dominate coastal weather.

Daytime High Tide:	*6:24 pm*	*7.1*
Daytime Low Tide:	*12:19 pm*	*3.4*
Sunrise:	*7:31 am*	
Sunset:	*4:35 pm*	

This long, cold, wet night was especially depressing. Every piece of canvas was now so rotted and torn it was useless to try to cover anything. Everyone, including the captains, was sleeping fully exposed to the rain.

> *Wind Shifted . . . and blew hard accompanied with hard rain all last night, we are all wet bedding and Stores, haveing nothing to keep our Selves or Stores dry* Moulton, *Journals* 6:91 (Clark)

They now found themselves in a situation that was almost identical to that which had occurred to them along the northern shore. They were trapped again by the waves and suffered for want of food. Every hunter came back empty-handed. This forced the men to continue eating the dried Indian fish, which though nourishing enough, was certainly not what anyone would want to eat three times a day, day after day.

> *the Swan and brant which are abundant Cannot be approached Sufficently near to be killed, and the wind and waves too high to proceed on to the place we expect to find Elk, & we have nothing to eate except pounded fish* Moulton, *Journals* 6:92 (Clark)

It occurred to one of the men that this unusual peninsula of land created a natural trap for animals. If most of the party walked out to the farthest tip of the peninsula and beat the bushes, all the deer would be forced to run directly towards the mainland. Then, if good marksmen stood at the narrow neck, which was only fifty yards wide, they could easily shoot the animals that came towards them.

> *we Sent out the most of the men to drive the point for deer, they Scattered through the point; Some Stood on the pensolu* Moulton, *Journals* 6:91 (Clark)

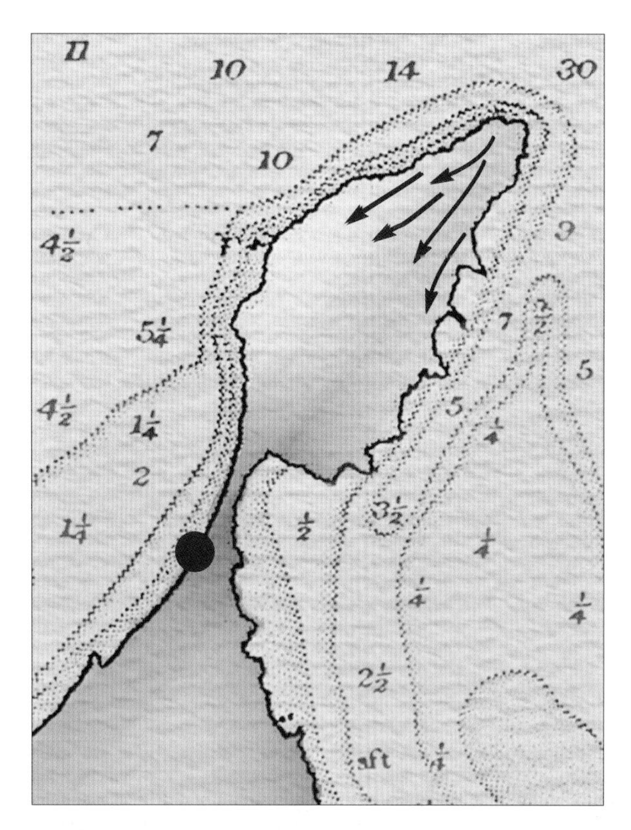

The captains position their best hunters near the narrow neck of land; the rest of the party drive all the wild game towards the ambush.

Animals of the rainforest blend in perfectly with their surroundings. Oftentimes they are practically impossible to see until they make a move.

This elaborate trap was carried out, but to the surprise and disappointment of the party, not a single animal was found.

we Could find no deer Moulton, *Journals* 6:91 (Clark)

Little by little, the reality of their situation sank in. The misery of cold and hunger made each passing minute feel like hours.

we have nothing to eate but a little Pounded fish which we purchasd. at the Great falls, This is our present Situation,! truly disagreeable. aded to this the robes of our Selves and men are all rotten from being Continually wet Moulton, *Journals* 6:91 (Clark)

To make matters worse, they couldn't turn around and go upriver even if they tried.

wind to high to go either back or forward

Moulton, *Journals* 6:91 (Clark)

The approaching storm hit them around noon. Lewis and Clark and their men hunkered down as tremendous, terrifying winds howled through their camp.

about 12 oClock the wind Shifted about to the N.W. and blew with great violence for the remainder of the day at maney times it blew for 15 or 20 minits with Such violence that I expected every moment to See trees taken up by the roots, Some were blown down Moulton, *Journals* 6:92 (Clark)

The coastal forests are accustomed to winds of eighty miles per hour. A few limbs may snap but usually the trees are unharmed. Clark's comment about trees being blown down is a strong indication that some wind gusts may have been much more powerful, very likely exceeding one hundred miles per hour.

Those Squals were Suckceeded by rain, !O how Tremendious is the day. This dredfull wind and rain Continued with intervales of fair weather, the greater part of the evening and night

Moulton, *Journals* 6:92 (Clark)

There is no mention in Clark's journal of what conversations may have occurred among the men, but it would be interesting to know what was said to John Shields. If the party had listened to Shields and Sacagawea, they would not have been in this awful predicament.

Hunters

At this moment in time Lewis and Clark needed the services of their very best hunters. Every member of the party had become practiced with a rifle; most were excellent shots. But the act of tracking down wild animals required more than good aim. Hunting was a special skill. Hunters knew the habits of wild animals, they knew where they slept and what they ate. In addition to this, good hunters were persistent, cunning, and stealthy.

Lewis and Clark had several men in their party who excelled as hunters. George Shannon, Francois Labiche, and John Collins were among the best; the brothers Joseph and Reuben Field were equally talented. But no one could match George Drouillard. This man was the most remarkable hunter of the entire party. Clark writes:

This morning Sent out Drewyer and one man to hunt, they returned in the evening Drewyer haveing killed 7 Elk; I scercely know how we Should Subsist, I beleive but badly if it was not for the exertions of this excellent hunter; maney others also exert themselves, but not being accquainted with the best method of finding and killing the elk . . . they are unsucksessfull in their exertions.[1]

Tides

Anyone who operates a boat in the lower Columbia River must pay close attention to the river's tides. This was as true for Lewis and Clark as it is today.

In the early phases of my research, in order to thoroughly understand Lewis and Clark's experience in the lower river, I found it necessary to have a tide table. For example, I knew they would be more inclined to move their canoes during *low water slack*, when the river flattens out smooth, or that an ebb tide in combination with a southwest wind would be very rough. Knowing the phase of the tide told me exactly what the river condition would be at each moment and that would help me predict their movements.

I hired National Oceanographic Atmosphere Administration engineers at Silver Springs, Mary land to *hind cast* back to November and December 1805. They made the computations and assured me that the data would be reasonably accurate.

I cross-checked Clark's journal and found his observations agreed with the NOAA data. For example, on December 6 Clark wrote:

> *High water to day at 12 oClock & 13 Inches higher than yesterday*[1]

Considering the regular fluctuations caused by wind, rain, and ocean surge, this table shows us that high tide did occur around noon, and was in fact an extremely high tide.

Tide Tables

Tide tables give three important bits of information about each tide. They tell us:

- when it will occur
- whether it will be high or low
- how much the water level will rise or drop

This information is represented in this form:

9:33 am H 8.5

This means that at around 9:30 in the morning the tide will be high and it will measure about 8.5 feet, which in the Columbia River is an average high tide.

7:53 pm L -1.5

This means a low tide will occur around 8 o'clock at night. The "minus" tide indicates a lower than average water level by 1.5 feet. Minus tides are extremely low, exposing more beach.

As a rule the tide changes every six or seven hours. The runoff from excessive rainfall can extend the ebb tide beyond its predicted time-table. On the other hand, a strong westerly wind pushing water into the river increases the predicted high tide.

THE COLUMBIA RIVER TIDE TABLE – 1805[2]

Calculated for Tongue Point

November

DATE	TIME	FEET	TIME	FEET	NOON	TIME	FEET	TIME	FEET
7 Th	1:24 am	H 7.2	6:53 am	L 2.4		12:53 pm	H 9.4	7:53 pm	L 1.4
8 Fr	2:11 am	H 7.2	7:35 am	L 2.7		1:33 pm	H 9.5	8:39 pm	L 1.5
9 Sa	3:00 am	H 7.1	8:21 am	L 2.9		2:17 pm	H 9.4	9:27 pm	L 1.4
10 Su	3:52 am	H 7.0	9:14 am	L 3.0		3:06 pm	H 9.1	10:18 pm	L 1.1
11 M	4:47 am	H 7.0	10:15am	L 3.2		4:01 pm	H 8.6	11:12 pm	L 0.7
12 Tu	5:47 am	H 7.0	11:25 am	L 3.2		5:06 pm	H 8.0		blank
13 W	12:09am	L 0.2	6:48 am	H 7.3		12:40 pm	L 2.9	6:22 pm	H 7.4
14 Th	1:08 am	L 0.3	7:47 am	H 7.6		1:56 pm	L 2.4	7:43 pm	H 6.9
15 F	2:08 am	L 0.8	8:24 am	H 8.1		3:06 pm	L 1.7	8:59 pm	H 6.7
16 Sa	3:06 am	L 1.2	9:33 am	H 8.5		4:08 pm	L 1.0	10:07 pm	H 6.8
17 Su	4:01 am	L 1.6	10:19 am	H 8.9		5:03 pm	L 0.3	11:08 pm	H 6.9
18 M	4:53 am	L 2.0	11:02 am	H 9.1		5:52 pm	L 0.3		blank
19 Tu	12:01 am	H 7.0	5:41 am	L 2.3		11:42 am	H 9.2	6:36 pm	L 0.6
20 W	12:49 am	H 7.1	6:25 am	L 2.6		12:20 pm	H 9.1	7:17 pm	L 0.8
21 Th	1:33 am	H 7.1	7:07 am	L 2.9		12:56 pm	H 9.0	7:56 pm	L 0.8
22 Fri	2:14am	H 7:1	7:46 am	L 3.1		1:30 pm	H 8.9	8:33 pm	L 0.7
23 Sa	2:54 am	H 7.0	8:23 am	L 3.3		2:04 pm	H 8.7	9:10 pm	L 0.5
24 Su	3:33 am	H 6.9	9:01 am	L 3.5		2:39 pm	H 8.4	9:48 pm	L 0.3
25 M	4:13 am	H 6.9	9:41am	L 3.6		3:16 pm	H 8.1	10:26 pm	L 0.0
26 Tu	4:55 am	H 6.9	10:26 am	L 3.6		3:58 pm	H 7.8	11:05 pm	L 0.4
27 W	5:39 am	H 7.0	11:19 am	L 3.6		4:47 pm	H 7.3	11:46 pm	L 0.8
28 Th	6:24 am	H 7.1	12:19 pm	L 3.4		5:47 pm	H 6.9		blank
29 F	12:31am	L 1.2	7:12 am	H 7.4		1:23 pm	L 3.0	6:58 pm	H 6.5
30 Sa	1:18 am	L 1.6	7:59 am	H 7.7		2:27 pm	L 2.4	8:14 pm	H 6.3

December

DATE	TIME	FEET	TIME	FEET	NOON	TIME	FEET	TIME	FEET
1 Su	2:09 am	L 2.1	8:45 am	H 8.1		3:27 pm	L 1.7	9:26 pm	H 6.3
2 M	3:02 am	L 2.5	9:30 am	H 8.5		4:23 pm	L 0.9	10:31 pm	H 6.4
3 Tu	3:56 am	L 2.8	10:15am	H 9.0		5:15 pm	L 0.1	11:30 pm	H 6.7
4 W	4:50 am	L 3.0	11:00 am	H 9.4		6:05 pm	L 0.6		blank
5 Th	12:24 am	H 7.0	5:42 am	L 3.2		11:45 am	H 9.7	6:53 pm	L 1.2
6 F	1:15 am	H 7.2	6:33 am	L 3.2		12:32 pm	H 9.9	7:41 pm	L 1.5
7 Sa	2:04 am	H 7.4	7:25 am	L 3.2		1:19 pm	H10.0	8:28 pm	L 1.6

CHAPTER EIGHT

Lewis's Excursion

where there is elk is our most eligible situation for a winter camp

Friday, November 29th

Overcast, rain showers continue. The jet stream moves up from the south.

Daytime High Tide:	*7:12 am*	*7.4*
Daytime Low Tide:	*1:23 pm*	*3.0*
Sunrise:	*7:33 am*	
Sunset:	*4:35 pm*	

The storm passed during the night, but the waters of the Columbia remained far too dangerous for Lewis and Clark to launch their heavy canoes.

> *Blew hard and rained the greater part of the last night and this morning, . . . The Swells and waves being too high for us to proceed down in our large Canoes, in Safty*
>
> Moulton, *Journals* 6:93 (Clark)

The men built bonfires to dry their tattered clothes so they could stitch the ragged pieces back together, but the blustery winds whipped the smoke in every direction, constantly interrupting their sewing.

> *The winds are from Such points that we cannot form our Camp So as to provent the Smoke which is emencely disagreeable, and painfull to the eyes*
>
> Moulton, *Journals* 6:93-94 (Clark)

Lewis and Clark could see rough water farther downriver, and they knew their large canoes could not endure such waves. So rather than let this delay them, the captains formed a new plan: they decided to split up the party. The majority of the men, with all the baggage and four large canoes, would stay where they were, and Captain Clark would remain with them. Meanwhile, a select group of their best hunters, led by Captain Lewis, would advance downriver. The waves of the Columbia were high, but the men were confident that their indispensable Indian canoe could navigate such waters safely.

As the hunters cleaned their rifles and sharpened their knives and tomahawks for this reconnaissance mission, Lewis decided to bring along something he had not touched for several months: a notebook and ink. This is worth mentioning for two reasons. First, Lewis had stopped writing when he became ill in September, and this resumption of his journal is a welcome return of his distinctive voice. Second, it gives us a priceless description of this surprising and dramatic search for elk.

Lewis begins his journal by describing where he is going and why. His words are sparse; he gets right to the point.

> *the wind being so high the party were unable to proceed with the perogues. I determined therefore to proceed down the river . . . set out early this morning in the small canoe accompanyed by 5 men*
>
> Moulton, *Journals* 6:92 (Lewis)

Lewis and his small party of hunters paddled downriver along the shore for three miles, rounded a point, and then entered a large bay near the Columbia's mouth. The western side of this bay, which was the place the Indians had

Lewis and five of his best hunters paddle downriver several miles and enter a large bay.

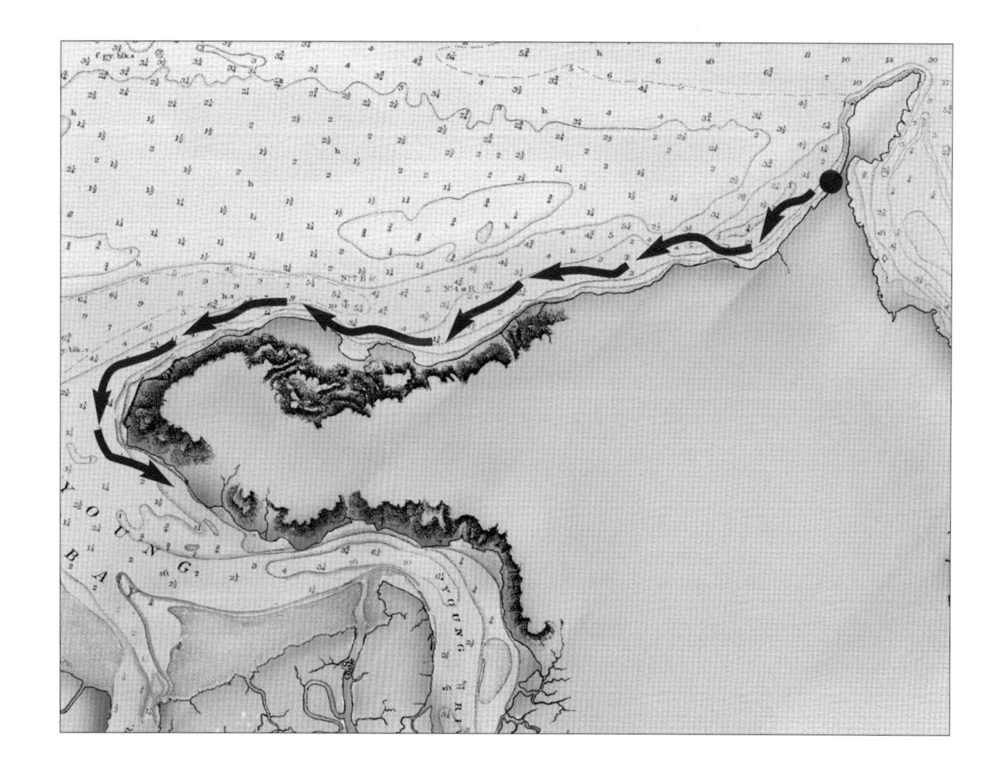

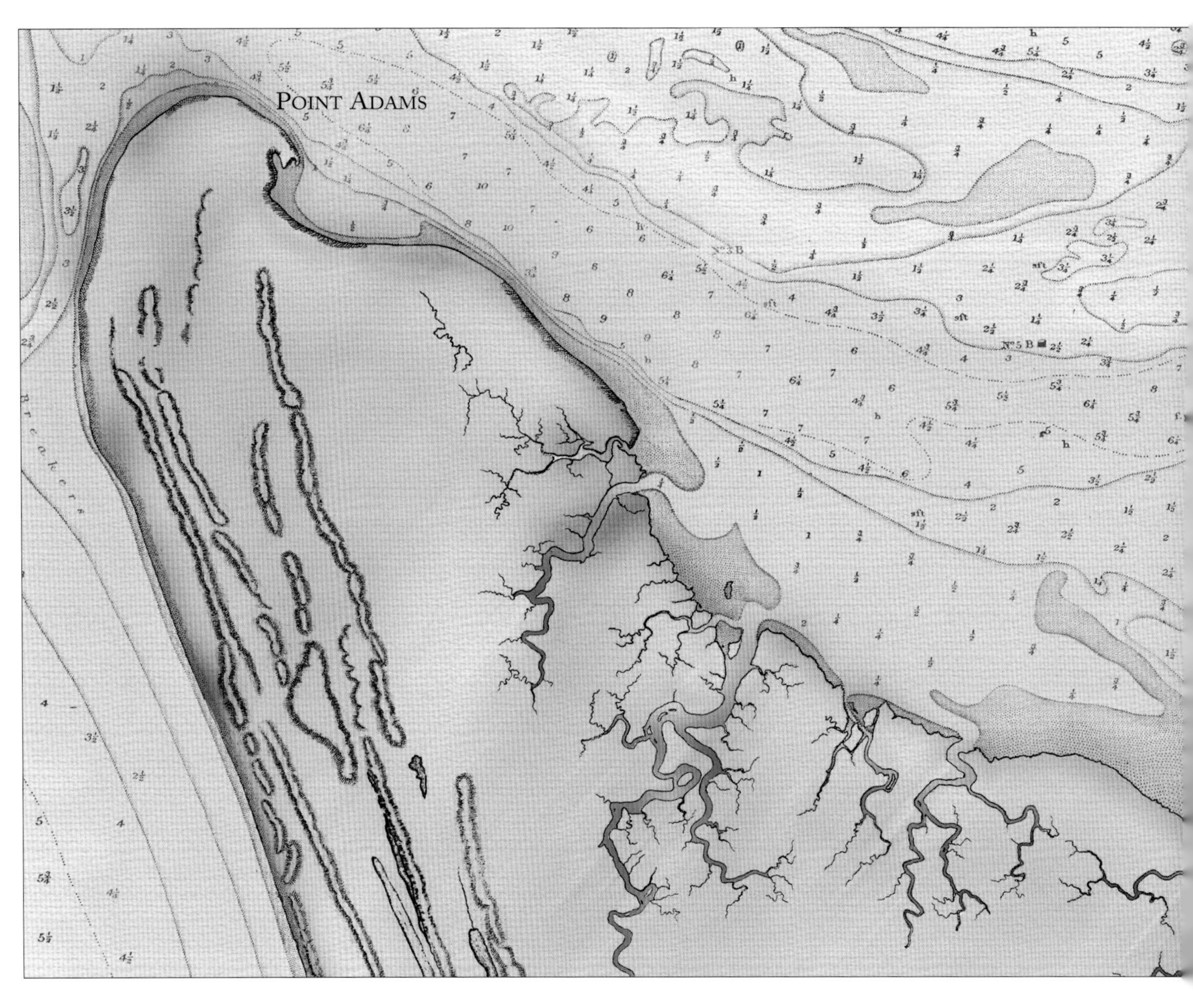

pointed out for good elk hunting, was still several miles away.[1] Realizing they could not reach there before dark, Lewis sent his men ashore to hunt at the nearest point. His intentions were to get food for that night and to check the availability of game.

The hunters had extraordinary success. They quickly returned with an abundance of game. The success of this single hunt must have thrilled Lewis and immediately convinced him that, exactly as the Clatsops had said, this area was filled with animals.

> *send out the hunters they killed 4 deer, 2 brant a goos and seven ducks* Moulton, *Journals* 6:93 (Lewis)

Lewis's party had done well. They had acquired enough food for dozens of people and had found an abandoned Indian hut where they could bivouac for the night. At this moment in time everything looked promising.

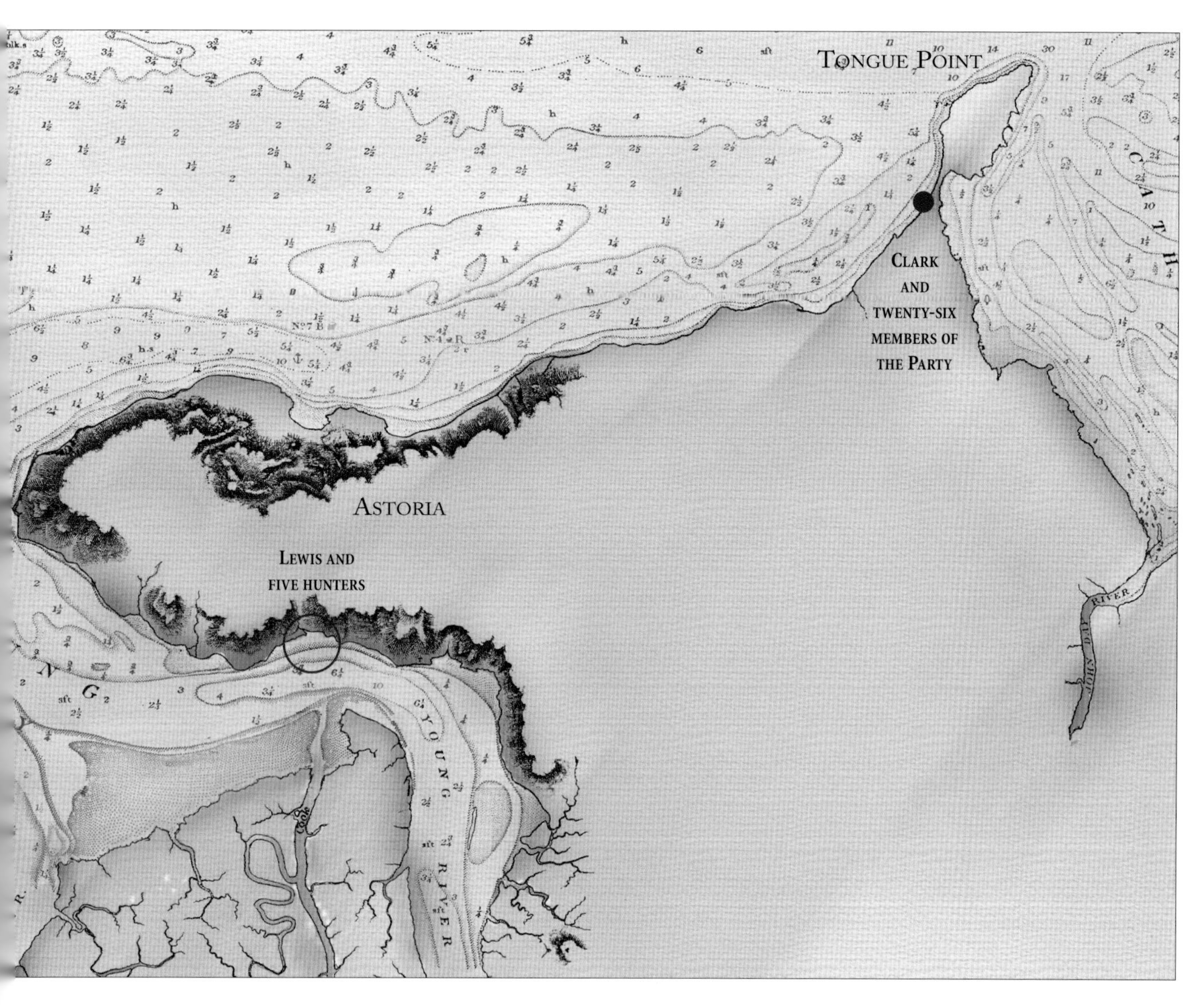

The party is split up into two separate groups. Clark and twenty-six members of the party remain near Tongue Point; Lewis and his hunting party camp along the shore of Youngs Bay.

Meanwhile at Clark's camp the situation was exactly the opposite. He had tried to provide his party with fresh game, but his hunters returned to camp with nothing. So Clark's men were forced to continue with the same diet as before.

> *our diat at this time and for Severall days past is the dried pounded fish we purchased at the falls boiled in a little Salt water* Moulton, *Journals* 6:93 (Clark)

Clark and his men were camped at a dreadful place. The shoreline faced directly downriver and was hit with every gust of wind. Fortunately, Lewis wasn't going far, and in a day or two he would return with a report of what he had found; then they could either move on or turn around and head upriver.

Saturday, November 30th

Hail and rain give way to calm, clear weather. High overcast skies with sunbreaks dominate the afternoon weather.

Daytime High Tide:	*7:59 am*	*7.7*
Daytime Low Tide:	*2:27 pm*	*2.4*
Sunrise:	*7:34 am*	
Sunset:	*4:34 pm*	

THE WEATHER WAS CLEAR AND CRISP - PERFECT for a hunt. Lewis and his party set out before dawn, paddled around the lip of the bay, and arrived in the vicinity of where the Clatsops had indicated that elk were abundant. Here the men found a small creek, which they followed deep into the interior of Point Adams.

we asscended this stream about 2 m.

Moulton, *Journals* 6:95 (Lewis)

Now it was time for the hunt to begin. Lewis sent several of his men westward, towards the ocean. For these hunters, tracking a herd of elk through swampy, muddy bottomland would be easy. The soft ground retains the impression of every hoof print; if there were an elk within two miles they would find it, track it down, and bring in the meat.

Sent out three men to examin the country to the S. & W.

Moulton, *Journals* 6:95 (Lewis)

Lewis could plainly hear the sound of the nearby ocean's surf; he could hear the cry of thousands and thousands of ducks and geese that swarmed overhead, but the sound Lewis most wanted to hear was the thunderous blast from one of his hunter's rifles. That sound would instantly tell him that his men had encountered a herd of elk.

He listened and waited. After several hours the men came back to camp bringing very bad news.

they returned after about 2 hours and informed me that the wood was so thick and obstructed by marrasses & lakes that they were unable to proceed

Moulton, *Journals* 6:95 (Lewis)

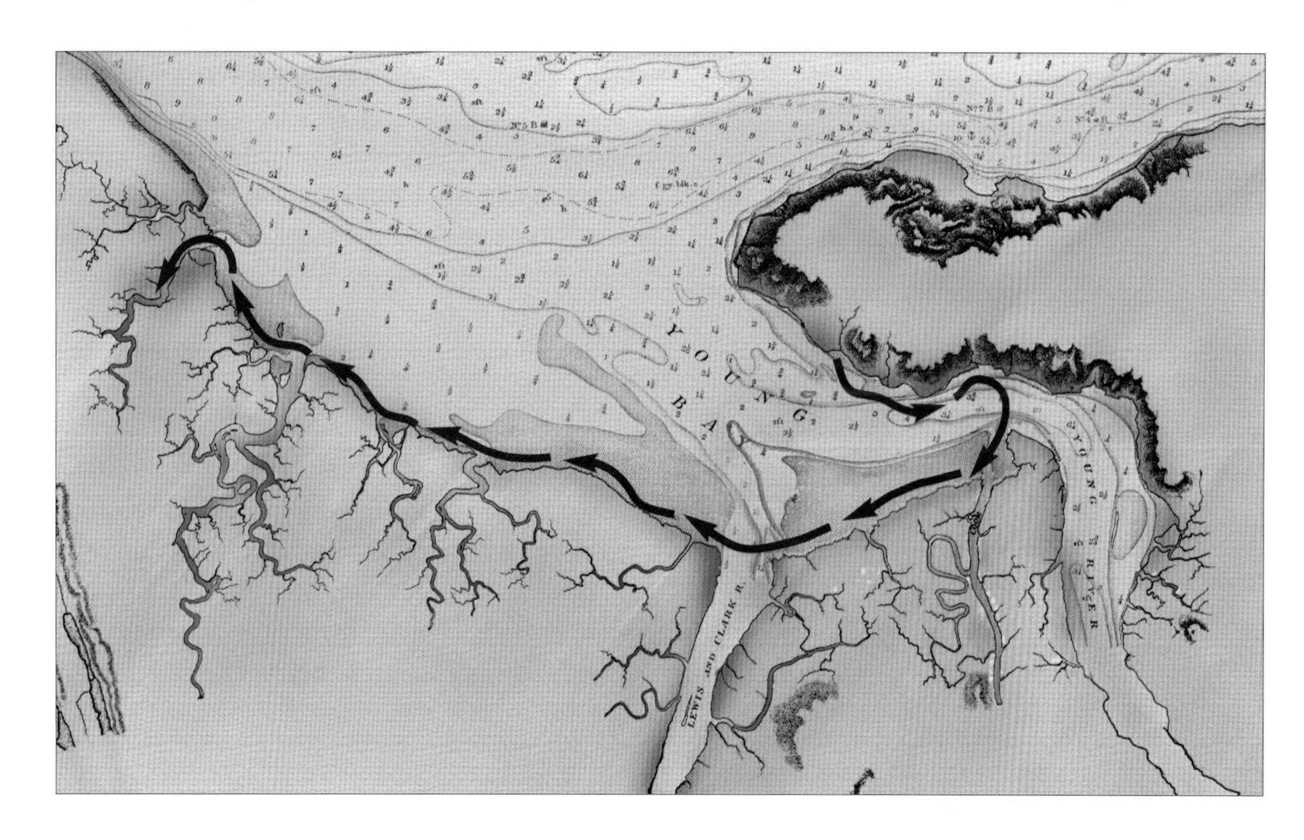

Lewis and his party paddle along the shore and enter a small stream where the Clatsops indicated elk were abundant.

This information no doubt surprised Lewis, but he immediately regrouped his men, paddled downstream to the bay, and set off to explore another inlet they had passed earlier that day. The Clatsops had pointed in this general vicinity, and Lewis probably assumed he had chosen the wrong stream.

> *we now returned and asscended the inlet which we had last passd* Moulton, *Journals* 6:95 (Lewis)

A good hunter uses his hearing to locate large animals in thick brush, so Lewis and his men paddled quietly up this creek, listening for sounds deep in the woods. Elk typically travel

Lewis does not draw maps and his descriptions are vague and confusing. It is impossible to know exactly which streams they ascended during this hunt. The routes shown here are speculative.

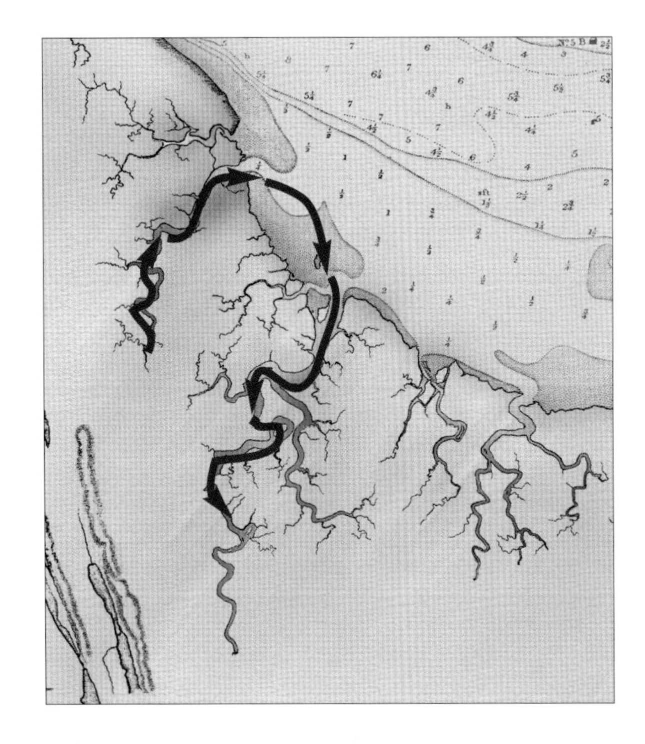

Finding no trace of game, they return to the bay and try a different nearby stream.

in large herds of thirty to fifty animals, and as they move through the thick brush, they can't avoid snapping limbs, which can be heard for a hundred yards.

Lewis's men were probably also watching the surface of the water. Elk are fearless swimmers, and they frequently plunge into streams. However, as they enter and leave the water their sharp hooves chop the vegetation along the shoreline, sprinkling grasses and leaves everywhere. When one sees bits of fresh-cut plants floating on the water, it's a sure bet that elk have recently crossed nearby. Try as they might, Lewis and his hunters found nothing.

> *no fresh appearance of Elk or deer in our rout so far. Asscend the inlet as we intended about 1 m. found it became much smaller* Moulton, *Journals* 6:95-96 (Lewis)

Lewis must have been completely perplexed. The Clatsop Indians had pointed directly at Point Adams and insisted that elk were abundant. However, from all indications, this southern side of the Columbia was exactly the same as the northern shore: the trees were the same, the wetlands were the same, and so was the complete absence of elk.

As Lewis looked around he must have been adding and multiplying numbers in his head. He knew his party could consume two elk each day, which meant that if they were to spend several months here they would need two hundred elk or more. He also knew that if this neighborhood were home to hundreds of large elk, they should have seen some signs of them by now; there should have been tracks, or trampled-down brush, or some scratch marks from antlers in the bark of trees. What he was seeing was not adding up.

Having seen no trace of animals, the men gave up the hunt, turned around, and returned to the bay. Lewis's patience had worn thin. Not only were there no elk, even the Clatsops had disappeared. There's an angry tone to his voice as he describes their predicament.

> *returned to the large arm of the bay which we passed this morning. here we expect to meet with*

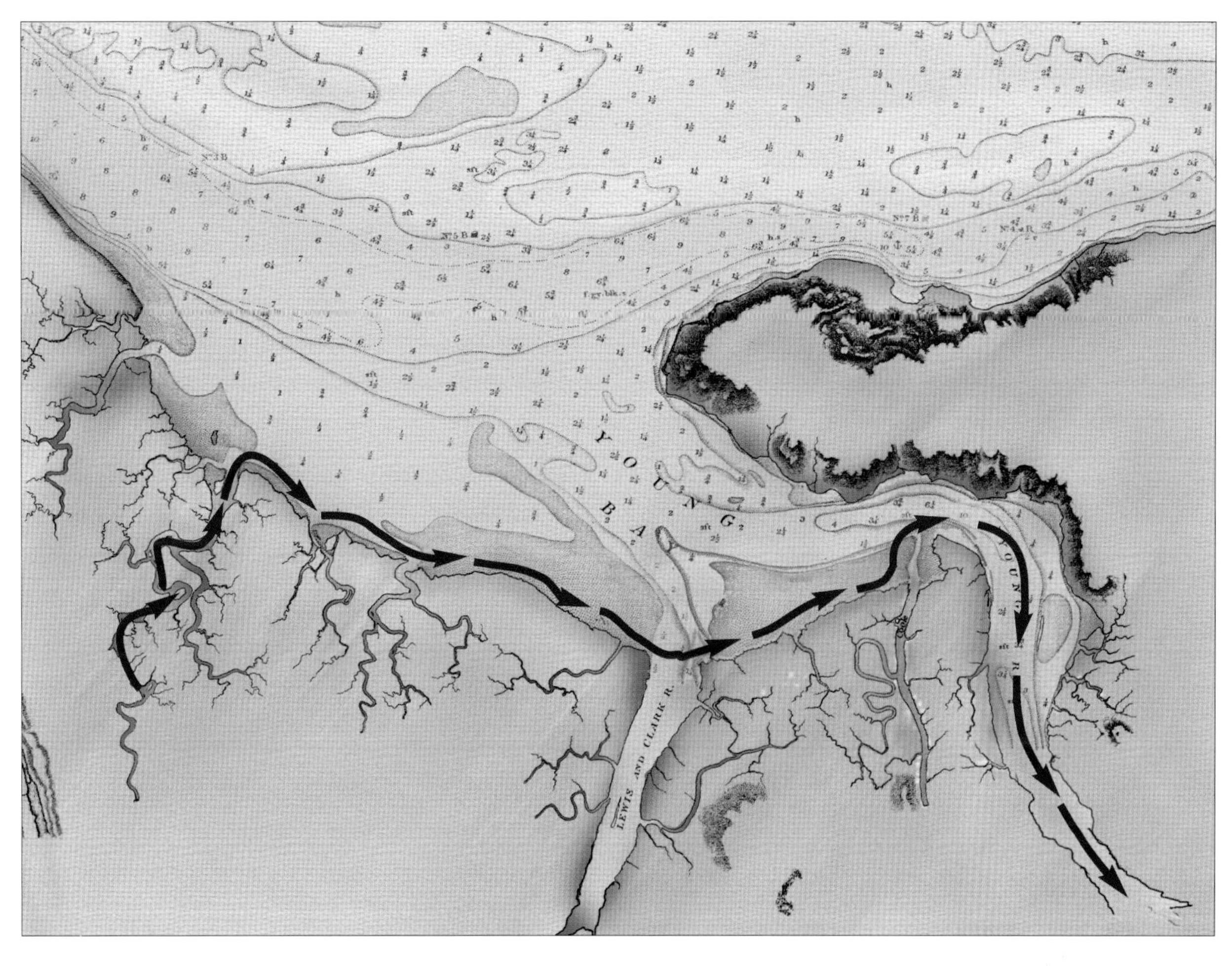

Finding no elk, Lewis and his hunters ascend a large tributary of the bay and set up camp.

> *the Clat-sop Indians, who have tantilized us with there being much game in their neighbourhood . . . for where there is most game is for us the most eligu-ble winter station* Moulton, *Journals* 6:95-96 (Lewis)

Lewis could have given up. He could have paddled directly back to Clark's camp, gathered the entire party together, and headed upriver to their second choice for a winter campsite. He would have described their hunt and how not a single elk track had been seen. His hunters would have confirmed this, and everyone in the party would have agreed to the change of plans.

Instead of giving up, however, Lewis remained determined. He decided to continue their search for elk in another location. This wasn't where the Indians had pointed, but maybe they would be lucky. His men turned their canoe into another tributary, paddled several miles, and camped.

> *continued our rout up the large arm of the bay . . . and encamped on the Stard. side on the highland*
> Moulton, *Journals* 6:96 (Lewis)

Clark's situation this day was bad and growing worse. The lack of fresh food was becoming a serious problem. All of his hunters returned empty-handed, which forced them to rely on the dried Indian fish; however, this was making his men ill. Clark searched for the reason and suspected it was their cooking water.

Several men Complain of a looseness and gripeing which I contribute to the diet, pounded fish mixed with Salt water, I derect that in future that the party mix the pounded fish with fresh water

Moulton, *Journals* 6:97 (Clark)

Unable to approach the alert swans and geese and equally frustrated by the absence of deer, some of Clark's hunters had turned their guns on a couple of unsavory coots and three hawks.[2] At this point, Clark and his men were so desperate they didn't care what kind of meat they ate.

my hunters killed three Hawks, which we found fat and delicious

Moulton, *Journals* 6:97 (Clark)

Young Sacagawea pitied Clark. It was one thing to be soaking wet and cold all day and night, but hunger was an entirely different problem and one that she understood. Sacagawea knew she had something that could relieve Captain Clark's suffering, so she unpacked a small portion of wheat bread that she had secretly carried for the past three months, and gave it to him.

The Squar gave me a piece of bread made of flour

Moulton, *Journals* 6:97 (Clark)

The young mother had carried this rock-hard crust of bread over the Rocky Mountains and for hundreds of miles by canoe, with the intention of giving it to her baby boy when he became old enough to chew food. Now, instead, she offered it to Clark.

She had reserved for her child and carefully Kept untill this time, which has unfortunately got wet, and a little Sour

Moulton, *Journals* 6:97 (Clark)

Clark could have refused it. He could have given it back to her for the baby, but he didn't. Instead, he immediately devoured every morsel and savored the taste of this unexpected treat.

this bread I eate with great Satisfaction, it being the only mouthfull I had tasted for Several months past

Moulton, *Journals* 6:97 (Clark)

As darkness approached, the parties were miles apart yet both practically in the same sad situation as they had been in the night before. However, some of Clark's hunters reported having had a glimpse of some elk, so there was a glimmer of hope that tomorrow they could pick up the track and make a kill.

"this bread I eate with great Satisfaction"

Lewis and Clark purchased tons of provisions for their party. They began their journey up the Missouri river with two tons of pork and ton of corn, plus beans, peas, biscuits, and coffee. Exactly how much wheat flour they brought along is unknown, but one list Clark made on the 14th of May, 1804, gives us a clue:

30 half Barrels of flour
3 Bags of [flour] weight 3400 lbs.[1]

They ate wild game whenever possible so that their provisions would last; however, in spite of their efforts, sixteen months later they were running low. The pork, beans, coffee, and biscuits were used up, and they had eaten the last of their wheat. Clark writes:

nothing to eate but berries, our flour out, and but little Corn, the hunters killed 2 pheasents only[2]

At some point, before their supply of flour had been depleted, Sacagawea set aside one piece of bread, carefully wrapped it, and kept it for her baby boy to eat when he became old enough for solid foods. She secretly carried this bread over the Bitterroot Mountains and down to the ocean, a distance of over six hundred miles.

Several types of raptors inhabit the lower Columbia River. Clark does not describe the birds his hunters shot, but red-tailed hawks are common and could be easily approached. The red-tails often perch in trees where they watch for any mice, snakes, rabbits, or squirrels that move into the open. They are quick and fearless hunters, but no match for a rifle.

Sunday, December 1st

Fair weather continues but a cold east wind blows down the Columbia.

Daytime High Tide:	*8:45 am*	*8.1*
Daytime Low Tide:	*3:27 pm*	*1.7*
Sunrise:	*7:36 am*	
Sunset:	*4:33 pm*	

Captain Lewis awakened to another perfect day. The weather was calm and the sky overcast. His hunters quickly disappeared into the underbrush, like a pack of hungry wolves.

> *Cloudy morning wind from the S.E. sent out the men to hunt* Moulton, *Journals* 6:101 (Lewis)

Meanwhile, Lewis remained at their campsite and waited. He kept a watchful eye on the canoe so that it wouldn't drift away and listened for the sound of a rifle.

This rare idle moment would have given Lewis an opportunity to pause and look at the unusual plants all around him. What he saw was unlike anything he had ever seen near his childhood home in Virginia, nor were these plants anything like those he had seen along the Missouri or in the Rocky Mountains. This was an entirely different plant community. These bushes, vines and shrubs were of a type that thrived where rainfall exceeded a hundred inches a year and the elevation was near sea level.

We have no idea how long Lewis waited for his men to return. Certainly they must have been gone for several hours, or possibly all morning. Eventually he heard the snap of a limb, a bit of rustling in the underbrush, and out stepped four of his hunters. Everyone, except for Drouillard, had returned, and they brought some very disappointing news.

> *they soon returned all except Drewyer and informed me that the wood was so thick it was almost impenetrable and that there was but little appearance of game; they had seen the track of one deer only and a few small grey squirrels* Moulton, *Journals* 6:101 (Lewis)

These hunters could find only one deer track. This news merely reinforced what they feared all along. It was now obvious: big game animals did not live anywhere near the ocean. The southern shore was just as barren as the northern shore.

What Lewis was thinking at this moment is, of course, unknown. He left no written description of his thoughts, but from his actions it is safe to assume that he now gave up on the idea of wintering here. He did not dispatch his hunters back out into the woods, nor did he paddle them farther upriver into new hunting territory. It was clear that this would be a waste of time; elk did not live around here.

The coastal rainforest.

As soon as his last hunter returned they would paddle back to Clark's camp, assemble the entire party, then continue on upriver. In the meantime, while he waited for Drouillard, Lewis picked up his pen and began to write a detailed and colorful description of the common Douglas pine squirrel his men had seen.

> *these suirrels are about the size of the red squirrel of . . . eastern Atlantic States, their bellies are of a redish yellow, . . . the tale flat and as long as the body . . . back and sides of a greyish brown*
>
> Moulton, *Journals* 6:101-102 (Lewis)

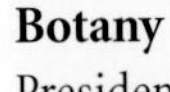

Botany

President Jefferson had requested that Lewis identify and collect any new plants he happened to encounter, but their arrival at the Pacific Ocean had been so challenging and the weather so wet that it was practically impossible to press and dry the leaves. Thus far, Lewis had collected only two plants while exploring around Cape Disappointment; these were the False Box (*Paschistime myrsinites*) and Feather Boa (*Egregia menziesii.).*[1]

He then turned his attention to a nearby bush and jotted down a description of this unusual plant.

> *the brier with a brown bark and three laves which put forth at the extremety of the twigs like the leaves of the blackbury brier, tho' is a kind of shrub and rises sometimes to the hight of 10 feet*
>
> Moulton, *Journals* 6:102-103 (Lewis)

It appears that Lewis suddenly awakened to the fact that he was surrounded by rare plants, none of which he had collected or described. Since he had now decided not to winter here,[3] he would not be able to include any of these plants in his botanical collection. However, he did have a window of opportunity at that moment. He and his men couldn't leave and rejoin Clark until Drouillard returned, so if he worked quickly, he could make a partial list of these unique coastal plants.[4] Without any more delay, he picked up his pen and began to write.

> *the green brier yet in leaf; the ash with a remarkable large leaf* Moulton, *Journals* 6:103 (Lewis)

There wasn't time for full descriptions. Lewis merely jotted down some key words which would be enough to jog his memory later. In a month or two, he would have the time to look over his notes and rewrite his descriptions in more detail.

> *the large black alder. the large elder with skey blue buries* Moulton, *Journals* 6:103 (Lewis)

Some plants were so unusual that Lewis was moved to describe them in great detail.

> *the broad leave shrub which grows something like the quill wood but has no joints, the leaf broad and deeply indented the bark peals hangs on the stem and is of a yelowish brown colour*
>
> Moulton, *Journals* 6:103 (Lewis)

Whenever possible, however, Lewis was thrifty with his words. He could describe three separate plants in fewer than twenty-five words.

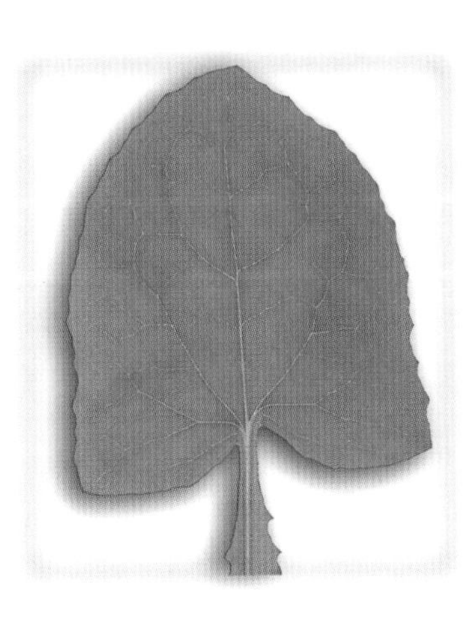

the seven bark is also found here as is the common low cramburry – there is a wild crab apple which the natives eat Moulton, *Journals* 6:103(Lewis)

Lewis does not record in his journal what the four hunters did after they returned from the hunt, but we do know what that fifth man was doing. The fifth man was George Drouillard, and he was doing exactly what Lewis had told him to do. He was hunting.

George Drouillard was an extraordinary man. He was clearly the party's most skillful hunter and woodsman. While Lewis sat describing the coastal plants in his notebook, this remarkable hunter slipped through the underbrush as quiet as a shadow, crossing streams without a splash. He pushed on deeper and deeper into the woods, always walking into the wind so that his scent wouldn't alert the animals. He stood motionless for minutes at a time, listening far into the thick under-story; he knelt down to examine bent blades of grass and overturned moss. Drouillard was relentless.

Meanwhile, Lewis was busy observing another plant.

the tree which bears a red burry in clusters of a round form and size of a red haw

Moulton, *Journals* 6:103 (Lewis)

When Lewis encountered the madroña tree he noted, quite accurately, its many unusual features.

the leaf like that of the small magnolia, and brark smoth and of a brickdust red coulour it appears to be of the evergreen kind Moulton, *Journals* 6:103 (Lewis)

Suddenly, a rifle blast shattered the silence. The echo of each thunderous shot had every man on his feet, looking around and trying to pinpoint where the sound had come from. Lewis's pen stopped moving, as he counted each time the trigger was pulled. Several minutes passed, then he jotted down the following words:

half after one oclock Drewyer not yet arrived. heard him shoot 5 times just above us and am in hopes he has fallen in with a gang of elk

Moulton, *Journals* 6:103 (Lewis)

After writing the words "a gang of elk" Lewis stopped writing. From this moment onward we have no idea of what he was thinking or doing. In fact, an entire month passed before he wrote another word.[5]

Lewis had no idea that those five shots had changed, once again, the entire course of their expedition.

Six miles away, the situation of Captain Clark and his men remained hopelessly unchanged. Hunger and the lack of fresh food was troubling everyone. In a desperate attempt to relieve this situation, Clark personally accompanied a group of hunters, perhaps hoping his presence would cause everyone to double their efforts. However, their canoe was battered by waves and tossed in such a manner as to force them to return to camp and to their same meager diet.

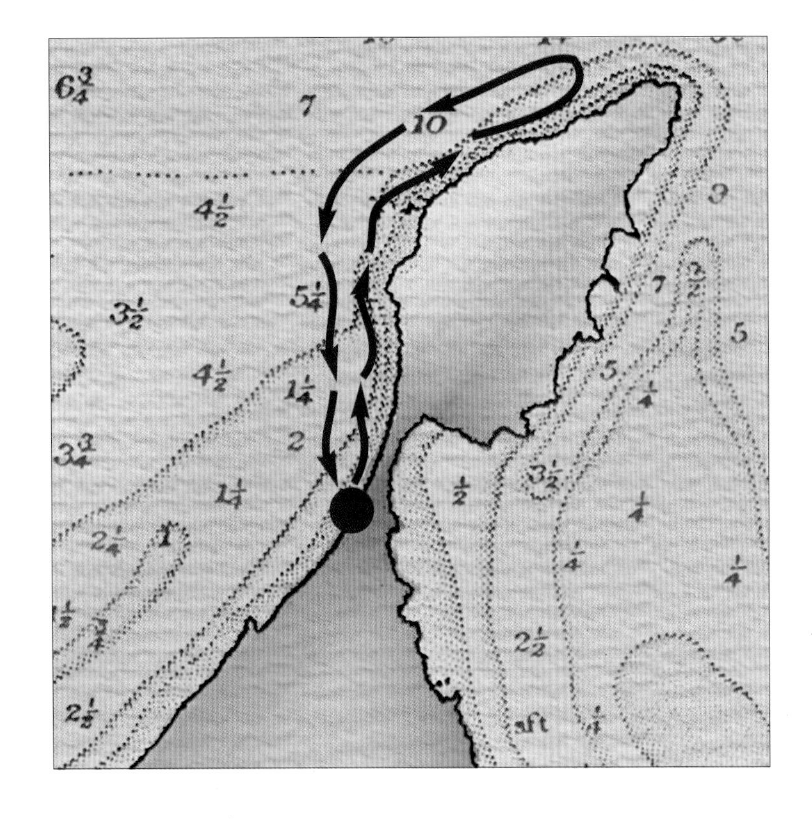

Clark attempts to go hunting, but the waves force his canoe to return to camp.

> *I deturmined to take a Canoe & a fiew men and hunt the marshey Islands . . . the Wind rose So high that I could not proceed, and returned to partake the dried fish, which is our Standing friend*
>
> Moulton, *Journals* 6:104 (Clark)

The men faced another long, miserable day. They dried their rotting buckskins by the fire and continued to stitch the pieces back together.

> *men all employed to day in mending their leather Clothes, Shoes, &c. and Dressing leather*
>
> Moulton, *Journals* 6:104 (Clark)

This was a difficult and depressing time for Captain Clark, and it could all be blamed on this particular stretch of violent river. Until now his men had dominated every obstacle nature had set before them: they had paddled against swift currents, outrun grizzly bears, portaged around waterfalls, and conquered every mountain. The mouth of this great river, however, humbled them. Clark described the view he had from this camp, looking directly into these powerful, unforgiving waters:

> *its waters are forming and . . . breake with emenc waves on the Sands and rockey Coasts, tempestous and horiable* Moulton, *Journals* 6:104 (Clark)

The noise created by the powerful current of the Columbia colliding head-on with the waters of the ocean reminded Clark of the sound of a waterfall.

> *The emence Seas and waves . . . roars like an emence fall at a distance* Moulton, *Journals* 6:104 (Clark)

Humor rarely appears in Clark's journal, but he cannot resist making a little joke out of the ocean's name. He thinks the word "pacific" is an entirely inappropriate name for this body of water.

> *and this roaring has continued ever Since our arrival . . . in Sight of the Great Western (for I cannot say Pacific) Ocian as I have not Seen one pacific day Since my arrival in its vicinity*
>
> Moulton, *Journals* 6:104 (Clark)

As evening passed into night, Clark's thoughts turned to Lewis. It had been three days since the hunting party had left. Where

were they? Lewis had gone only a few miles away. He and his men either should have found some elk or returned by now.

I have no account of Capt. Lewis Since he left me

Moulton, *Journals* 6:104 (Clark)

Night came, and once again the men settled in for another long wait until sunrise. Perhaps Lewis would return tomorrow.

The roar of the ocean during winter gales can be heard inland for twenty miles.

CHAPTER NINE

Clark Waits for Lewis

1000 conjectures
has crowded into my mind

Monday, December 2nd

Rain showers return to the coast.

Daytime High Tide:	*9:30 am*	*8.5*
Daytime Low Tide:	*4:23 pm*	*0.9*
Sunrise:	*7:37 am*	
Sunset:	*4:33 pm*	

The repetitive diet of dried fish had finally taken its toll. Cramps, indigestion, and diarrhea sickened many of Clark's men. Worse yet, even the captain himself awakened feeling ill.

> *I feel verry unwell, and have entirely lost my appetite for the Dried pounded fish which is in fact the cause of my disorder at present – The men are generally Complaining of a lax and gripeing*
>
> Moulton, *Journals* 6:105 (Clark)

Clark was now faced with a terrible dilemma. The lack of fresh food was causing his men to fall ill, and the only solution was to move their camp closer to good hunting. However, loading and launching the canoes would be strenuous, and handling them in these terrible waves required tremendous effort and exertion from everyone. How could they do this if they were sick?

Another problem Clark faced was that, despite his illness, he had to be in charge and make decisions, although that was the last thing he felt like doing. Naturally, his thoughts turned to Lewis. Clark needed him here more than ever, to take charge of the party until he recovered his strength.

> *I expect Capt. Lewis will return to day with the hunters and let us know if Elk or deer Can be found Sufficent for us to winter on*
>
> Moulton, *Journals* 6:105 (Clark)

Four days had passed since Lewis had left to go hunting; if he didn't return soon, Clark would be forced to move their camp without him. It certainly would not be easy to do, but he had no other choice. He had to get his men closer to where there was food.

> *If he does not come I Shall move from this place, to one of better prospects for game*
>
> Moulton, *Journals* 6:105 (Clark)

Clark continued to send out more squads of hunters and some fishermen. All he could do for the moment was to keep trying to find some fresh food for his party.

> *I despatched 3 men to hunt and 2 and my Servent in a Canoe to a Creek above to try & Catch Some fish*
>
> Moulton, *Journals* 6:105 (Clark)

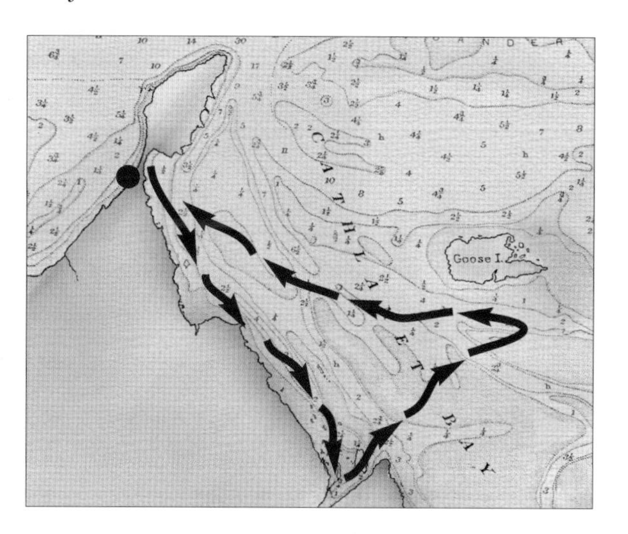

Clark sends fishermen to a nearby stream in a desperate attempt to find food. They fail. On their return the party tries to shoot some ducks or geese, but they fly off before the men can get within range.

The fishermen failed miserably. On their way back to camp they took a side trip, attempting to shoot some ducks, but could not even get close. The three elk hunters, however, fared better. They split into two separate parties so they could cover more territory. Sergeant Pryor and Gibson headed west and Joseph Field went east.

We lose track of Pryor and Gibson; however, Joseph Field's hunt was well documented, and the ordeal this remarkable man put himself through deserves recognition. He tracked down one herd of elk and crawled within range, but the pouring rain prevented his rifle from firing. Flintlocks were not designed for use in a rainforest, especially in the middle of winter.

Field shook the wet powder out of his rifle, reloaded, and continued on. Eventually he found another herd, crept up within range, but once again his rifle misfired. He must have been heartsick, but rather than give up, Field reloaded once again, and pushed on deeper and deeper into the dark, damp rainforest. His determination perhaps indicates how truly desperate their situation had become.

Clark sends out two squads of hunters. Sgt. Pryor and Gibson head west; Joseph Field goes east.

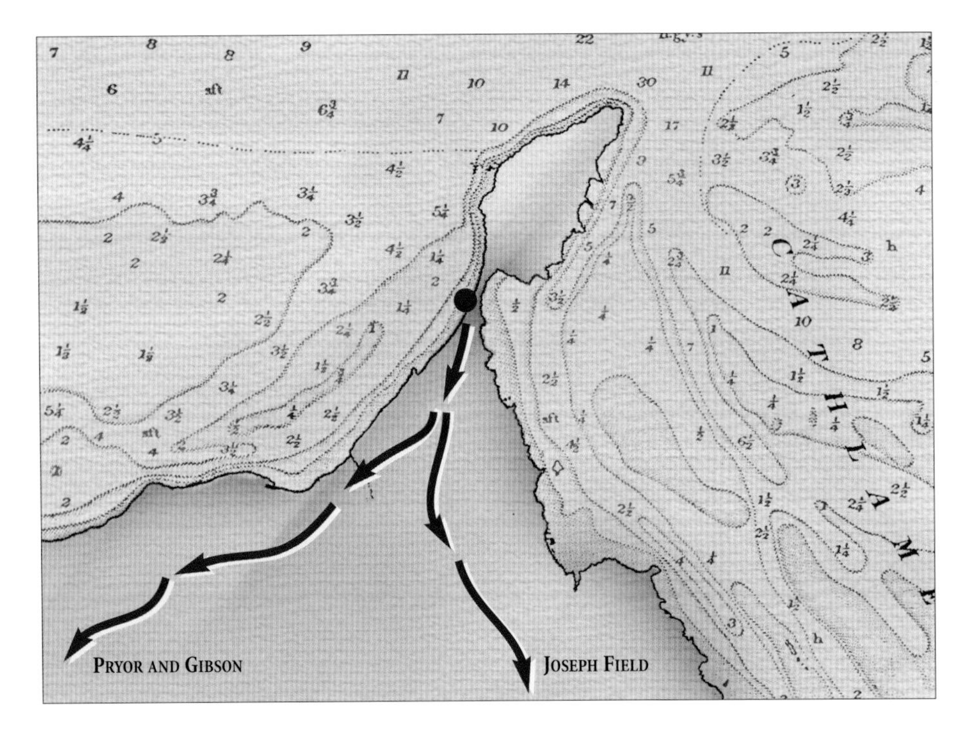

Finally, Field crept up within range of another herd, lined up his rifle, and pulled the trigger. Smoke and sparks erupted in one instantaneous flash; this time the powder ignited, the rifle recoiled with a thump against his shoulder, and a hundred yards away one elk fell dead.

Now Field was faced with a problem. He had chased these elk so far into the woods that he was at least half a dozen miles away from their camp. If he tried to carry a large load of meat by himself, he might not arrive before dark. His best move was to hurry back and get help.

However, his own deep pangs of hunger would have reminded him how desperate his companions were for any morsel of food. So, before leaving the kill, he drew his knife and cut out the huge leg bones of the elk. These he quickly tied into a bundle and slung over his shoulder. Carrying a few bones wouldn't slow him down.

Joseph hurried over fallen logs, pushed through thick brush, and sloshed across streams. His wet buckskin clothes clung to his body like a plaster of Paris cast and made every step an effort. It was an exhausting pace, but this remarkable woodsman moved so quickly that he reached camp before dark. Clark described his arrival.

> *In the evening Joseph Field came in with the Marrow bones of a elk which he killed at 6 miles distant, this welcome news to us*
>
> Moulton, *Journals* 6:105 (Clark)

Clark immediately assembled a party of men to retrieve the meat. No one except Joseph knew the location of the elk, so he had to turn around

It is easy to lose one's way when chasing elk through the dense coastal rainforest.

On this night the party is split up into four separate groups. Clark and seventeen others are at Tongue Point; Joseph Field and six men were caught in the dark while retrieving elk meat. Sgt Pryor and Gibson are somewhere on the back side of Astoria, and Lewis with his five hunters are around Youngs Bay.

and lead the party back out into the woods. This exhausting effort by Field, half starved, cold, and soaking wet, bordered on heroism.

> *I dispatched Six men in a empty Canoe with Jo; mediately for the elk which he Said was about 3 miles from the water* Moulton, *Journals* 6:105 (Clark)

Clark's men were undoubtedly elated by Field's unexpected gift. They smashed open each bone, scooped out every trace of marrow, and probably boiled it with dried fish to make a soup. The marrow from several leg bones divided among seventeen hungry stomachs would not have ended their misery, but it did give them a taste of what was coming. As soon as Joseph Field and the others returned with that canoe-load of meat, they would have the feast of their dreams.

> *this is the first Elk which had been killd. on this Side of the rockey mountains* Moulton, *Journals* 6:105 (Clark)

The men's spirits were lifted not only by the thought of the elk, but also by the news that Field had given Clark. Contrary to their first impressions of this area, it now appeared as though elk were indeed everywhere.

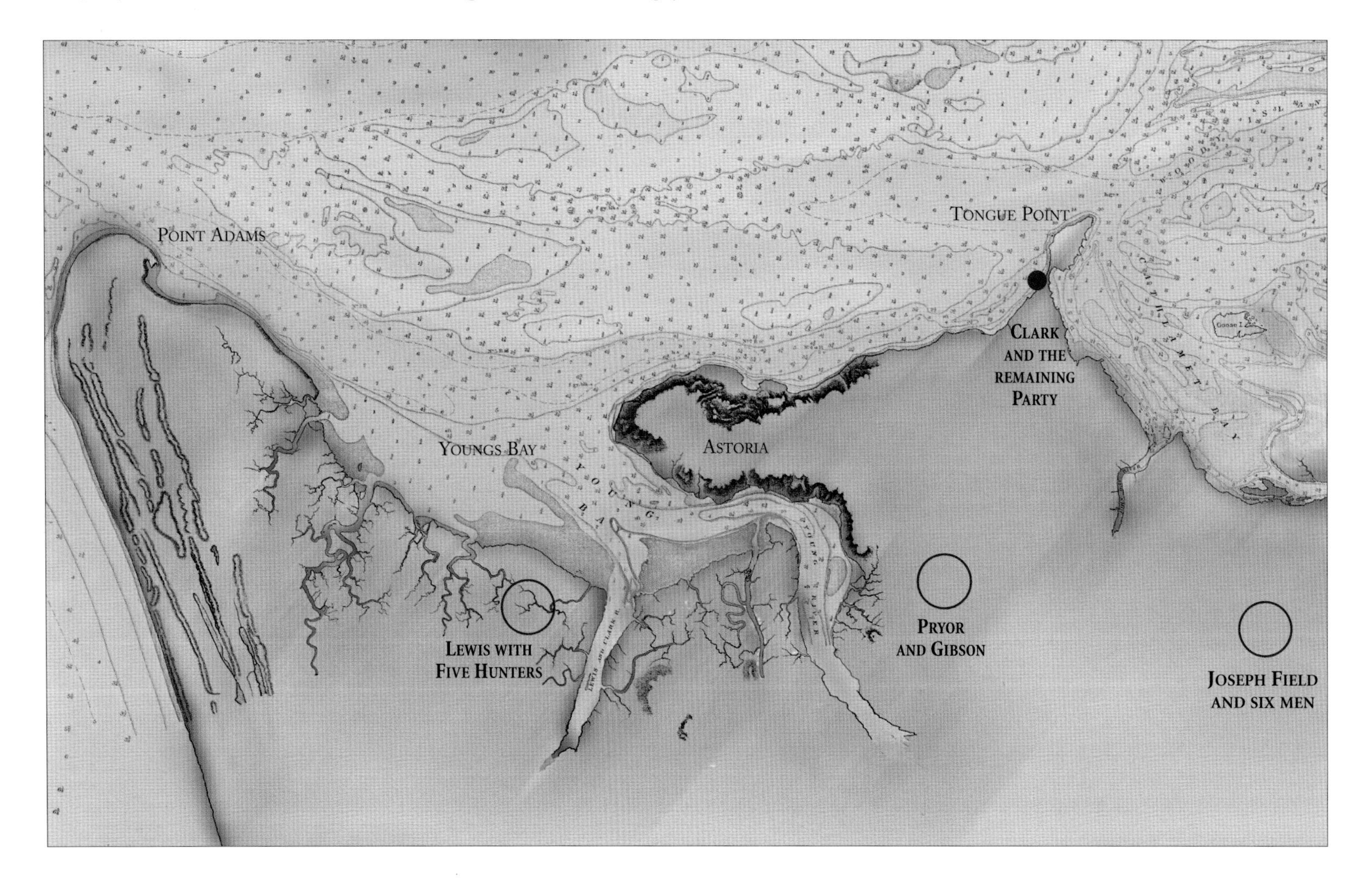

> *Jo Fields givs me an account of a great deel of Elk Sign & Says he Saw 2 Gangs of those Animals in his rout, but it rained So hard that he could not Shoot them*
>
> Moulton, *Journals* 6:105 (Clark)

Things were looking much better for Clark and his men. It could be said that Joseph Field had saved the day. The men waited and watched for his party to bring in the elk, hoping they would return that same night, but they didn't.

We have no written account of where Lewis was or what he was doing at this time. From Clark's point of view, Lewis and his five hunters had simply disappeared.

The mature elk that inhabit the Pacific Northwest can range in weight from six to eight hundred pounds. We have no record of how large an elk Field shoots, but the fact that six men were sent to retrieve the meat perhaps indicates it is a very large animal.

Tuesday, December 3rd

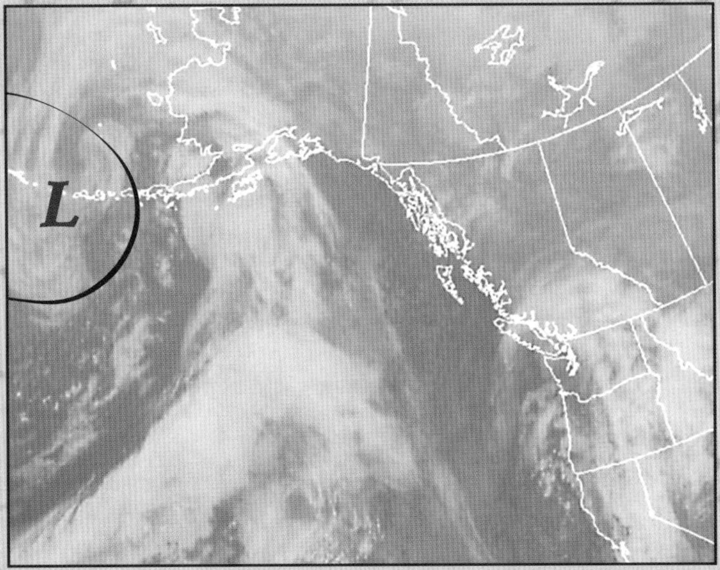

Fair and windy. A small low pressure system approaches the coast once again.

Daytime High Tide:	*10:15 am*	*9.0*
Daytime Low Tide:	*5:15 pm*	*0.1*
Sunrise:	*7:39 am*	
Sunset:	*4:32 pm*	

CLARK'S CAMP MUST HAVE BEEN A BEEHIVE OF excitement as the men eagerly awaited the returning canoe. Cooking fires were probably ablaze and kettles boiling in anticipation of the fresh meat. When Joseph Field and his squad finally touched shore in front of the camp, the men must have pounced upon the elk like pirates seizing a treasure ship.

> *the men Sent after an Elk yesterday returnd. with an Elk which revived the Sperits of my men verry much*
> Moulton, *Journals* 6:106 (Clark)

The following joyous hours of eating would have been a wonderful scene to watch; and that, unfortunately, was all that Captain Clark could do. While his men were stuffing themselves full of roasted elk, Clark's stomach would not allow him to touch a single bite.

> *I am unwell and cannot Eate, the flesh O! how disagreeable my Situation, a plenty of meat and incaple of eateing any*
> Moulton, *Journals* 6:106 (Clark)

Fortunately, some Indians passed downriver with a load of wappatoe roots,[1] which Clark was able to purchase. The wild tubers were boiled with elk meat into a soup.

> *those roots I eate with a little Elks Soupe which I found gave me great relief I found the roots both nurishing and as a check to my disorder*
> Moulton, *Journals* 6:106 (Clark)

We can only speculate how this elk meat was divided among the party. It would have made sense for Clark to ration it out among the men so it would last for several meals, yet there is one indication that all the meat was consumed before the day was out. Clark tells us that Sacagawea knew a technique for extracting every ounce of fat and oil from bones, which she probably would not have done if there had been a surplus of food.

> *after eateing the marrow out of two Shank bones of an Elk, the Squar choped the bones fine boiled them and extracted a pint of Grease, which is Superior to the tallow of the animal*
> Moulton, *Journals* 6:107 (Clark)

Like a pride of lions, the men stuffed themselves, then lay around digesting their meal.

Hungry Men eat elk in two meals

When meat was readily available, Lewis and Clark's men were capable of eating an unimaginably large quantity. Lewis makes this observation in July 1805, as the party ascended the Missouri.

> *we eat an emensity of meat; it requires 4 deer, an Elk and a deer, or one buffaloe to supply us plentifully 24 hours* [1]

On January 6, 1806, Clark and twelve men stumble upon a herd of elk. Clark writes:

> *I divided the party So as to be Certain of an elk, Several Shot were fired only one Elk fell, I had this Elk butchered and carried to a Creak in advance at which place I intended to encamp* [2]

The next morning Clark continues:

> *I[t] may appear Somewhat incrediable, but So it is that the Elk which was killed last evening was eaten except about 8 pounds* [3]

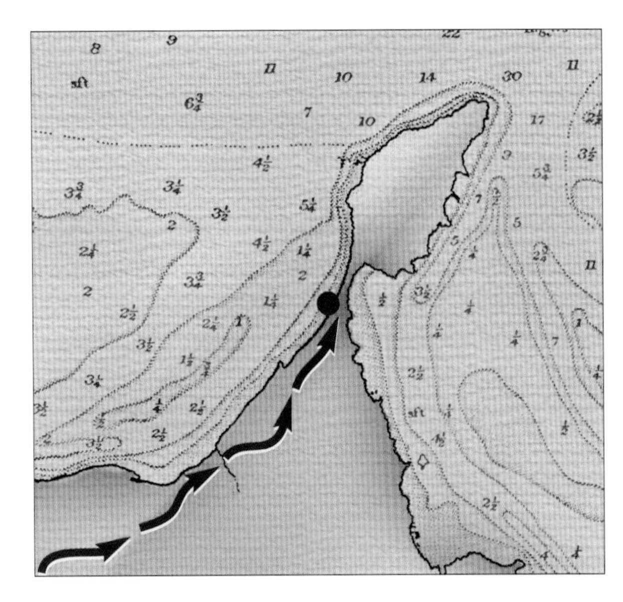

Sgt. Pryor and Gibson return from their elk hunt with good news.

This was the day to recover their health and strength. As darkness fell over the coastline, the men's quiet evening was unexpectedly interrupted when two more hunters wandered back to camp with some interesting stories.

> *Serjt. Pryor & Gibson who went hunting yesterday has not returned untill after night*
>
> Moulton, *Journals* 6:106 (Clark)

The two men had become lost while hiking through the labyrinth of old-growth rainforest. The trees were so thick and the forest canopy so dense and dark that they became entirely disoriented and wandered around completely confused.

> *informed me they had been lost the greater part of the time they were out* Moulton, *Journals* 6:106 (Clark)

Despite their embarrassing ordeal, Pryor and Gibson brought back some excellent news. During their hunt they had jumped a herd of elk and made several kills. The meat remained out in the woods, but as soon as they brought it into camp the feast would continue for several more days.

> *they informed me that they had killed 6 Elk at a great distance which they left lying, haveing taken out their interals . . . in their ramble saw a great deel of Elk Sign* Moulton, *Journals* 6:106-107 (Clark)

This had been the best day in a long, long time. The men's stomachs were stuffed full of meat, and their future looked bright.

Lewis's whereabouts, however, remained unknown. Five days had now passed since he had departed, five days without any word. Clark grew more worried with every passing hour. Something was obviously very, very wrong.

Some of Lewis and Clark's men become confused when hiking through the rainforest. They were unfamiliar with the terrain and vegetation; they were chasing elk that led them deep into the woods.

Wednesday, December 4th

Rain continues with blustery wind.

Daytime High Tide:	*11:00 am*	*9.4*
Daytime Low Tide:	*6:05 pm*	*–0.6*
Sunrise:	*7:40 am*	
Sunset:	*4:31 pm*	

During the night Clark recovered his strength. The wappatoe soup he had eaten had worked like a powerful medicine.

> *my appetite has returned and I feel much better of my late complaint* Moulton, *Journals* 6:107 (Clark)

However, Lewis's absence was obviously weighing heavily on Clark's mind. This was now the sixth day since Lewis and his men had gone ahead scouting for elk. Clark suspected the worst:

> *no account of Capt. Lewis. I fear Some accident has taken place in his craft or party*
>
> Moulton, *Journals* 6:108 (Clark)

Clark now faced a dilemma. He wanted to search for Lewis, yet his men desperately needed the elk meat his hunters had shot. The best solution to both problems was to move the camp downriver, closer to where Lewis had gone, and simultaneously have Pryor's men bring the meat directly there.

> *I Set out Sergiant Pryor and 6 men to the Elk he had killed with directions to Carry the meat to a bay which he informed me was below and as he believed at no great distance from the Elk, and I Should proceed on to that bay as Soon as the wind would lay a little and the tide went out in the evening* Moulton, *Journals* 6:107 (Clark)

Sergeant Pryor and his party set out while the others packed up their camping gear and loaded the four canoes. Clark was hoping the river would flatten; he was gambling on his usual good luck, but the mouth of the river refused to cooperate.

> *Hard wind from the South this evening, rained moderately all day and the waves too high for me to proceed in Safty to the bay as I intended*
>
> Moulton, *Journals* 6:107-108 (Clark)

The men sat and waited all through the middle of the day and into the late afternoon. The river roared and tossed with whitecapped waves. Clark must have been beside himself with frustration. Nothing was going as he had planned.

Clark plans to move the party downriver into a new camp where they could begin to search for Lewis. Sgt. Pryor is ordered to carry the elk meat to the bay and meet him there.

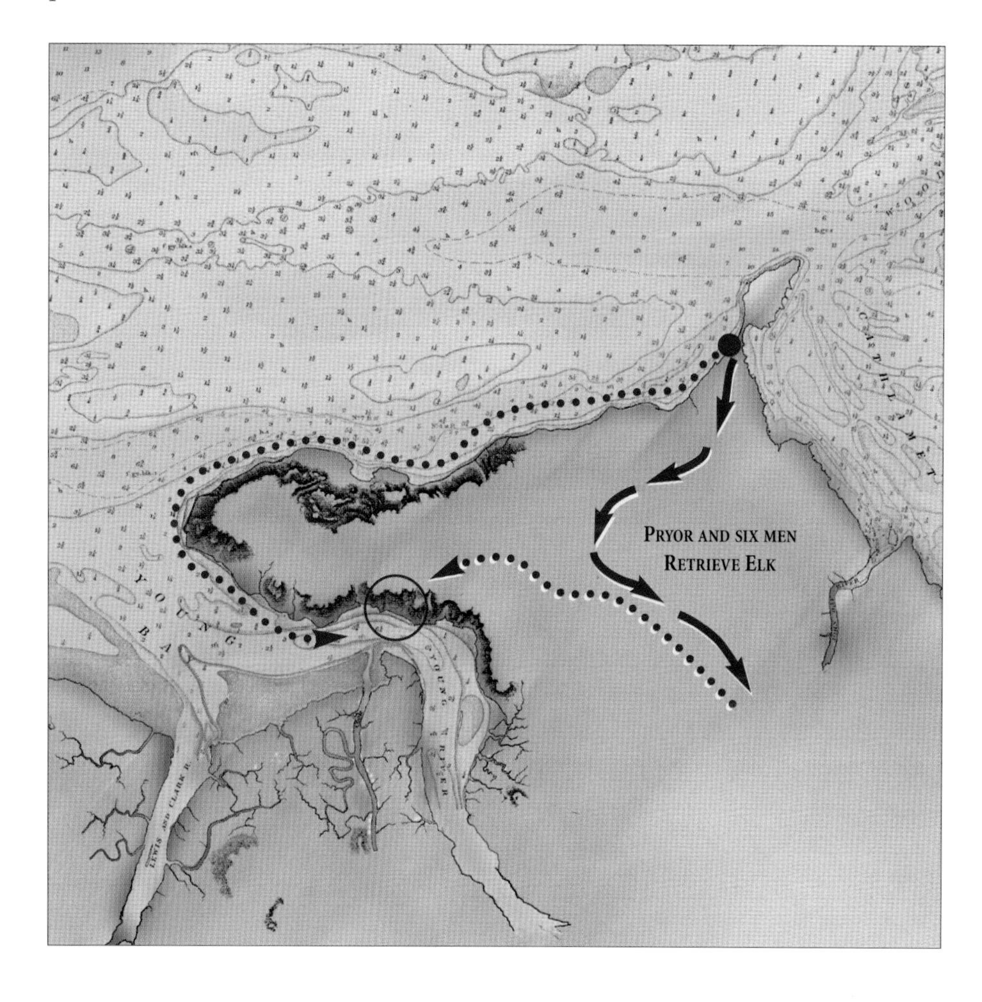

Thursday, December 5th

Rain continues with strong wind gusts from the south. A small low pressure system approaches the coast.

Daytime Low Tide:	*5:42 am*	*3.2*
Daytime High Tide:	*11:45 pm*	*9.7*
Sunrise:	*7:41am*	
Sunset:	*4:31 pm*	

Some hard showers of rain last night, this morning Cloudy and drisley at Some distance above the
rain is much harder high water to day at 12
this tide is 2 [illegible] higher [illegible] that [illegible]
all our Stores and bedding are again wet by
the hard rain of last night — Capt Lewis long
delay below, has been [illegible]
uneasiness on my part [illegible]
and Safty the [illegible]
which blows from the [illegible]
for me to know with [illegible]
an unknown Coast [illegible]
dis agreeable, the party much better of
indispositions — Capt Lewis [illegible] 3
men in the Canoe and informs me that he
thinks that a sufficient number of Elk may
be procured convenient to a situation on a
small river which [illegible] small bay
a short distance [illegible] party
had killed 6 Elk & 5 deer in his rout two

TORRENTS OF RAIN POURED DOWN WITHOUT A pause. The night was pure misery.

all our Stores and bedding are again wet by the hard rain of last night

Moulton, *Journals* 6:108 (Clark)

On top of this, the river was still not yielding a single inch.

the repeeted rains and hard winds . . . renders it impossible for me to move with loaded Canoes along an unknown Coast we are all wet & disagreeable

Moulton, *Journals* 6:108 (Clark)

Clark's thoughts remained with Lewis. There was no reasonable excuse for him to be gone an entire week. He should have found elk on his first day of hunting and returned. It now seemed obvious that something terrible had happened to him.

Capt Lewis's long delay below has been the cause of no little uneasiness on my part for him

Moulton, *Journals* 6:108 (Clark)

Perhaps some Indians had lured Lewis's party into an ambush, or maybe their unattended canoe had floated off with the tide and left them stranded. A gigantic rogue wave might have blindsided them, shattering their canoe against the rocks, or perhaps they had wandered too far into the dense woods and had become hopelessly lost. There were many, many ways to run into trouble in this dangerous, inhospitable, wild country.

a 1000 conjectures has crouded into my mind respecting his probable Situation & Safty

Moulton, *Journals* 6:108 (Clark)

Clark and his men grimly watched the river. They were prepared to make a dash downriver with their canoes the moment the waves grew calm.

As the men were waiting they would have noticed in the far, far distance four Indians advancing along the shore towards them. Maybe they were the same Indians who had sold Clark the wappatoe; maybe they would have some valuable news. As their canoe drew closer, Clark made a thrilling discovery: these were not Indians, it was Captain Lewis! Two of his men were missing but he had returned alive![2]

Rain!

Clark must have been elated to see Lewis returning to camp, but one would never know it from reading his journal. The thousand

Storm clouds, waves, rain – dark, threatening, distressful weather.

conjectures that had crowded his mind, the days of anxiety and worry, all disappeared. He regained his composure and wrote the following unemotional, matter-of-fact sentence:

> *Capt. Lewis returned haveing found a good Situation and Elk Suffient to winter on*
>
> Moulton, *Journals* 6:108 (Clark)

Clark's brief statement does not explain where Lewis had been for the preceding week, but it is easy to look back and piece the story together.

The five rifle shots Lewis had heard a week earlier had indeed been from Drouillard, and this extraordinary hunter had discovered how to find the well-camouflaged, elusive elk in the thick, coastal rainforest. However, finding elk was just the beginning of the work to be done.

Lewis knew it made absolutely no sense to even think of moving the party unless they could be sure that there were enough elk to last an entire winter.[3] So his first order of business was to have his men take a census of the region's elk population. How many days this took is anyone's guess, but they would have had to confirm that at least two or three hundred elk were close by before they could feel comfortable wintering there.

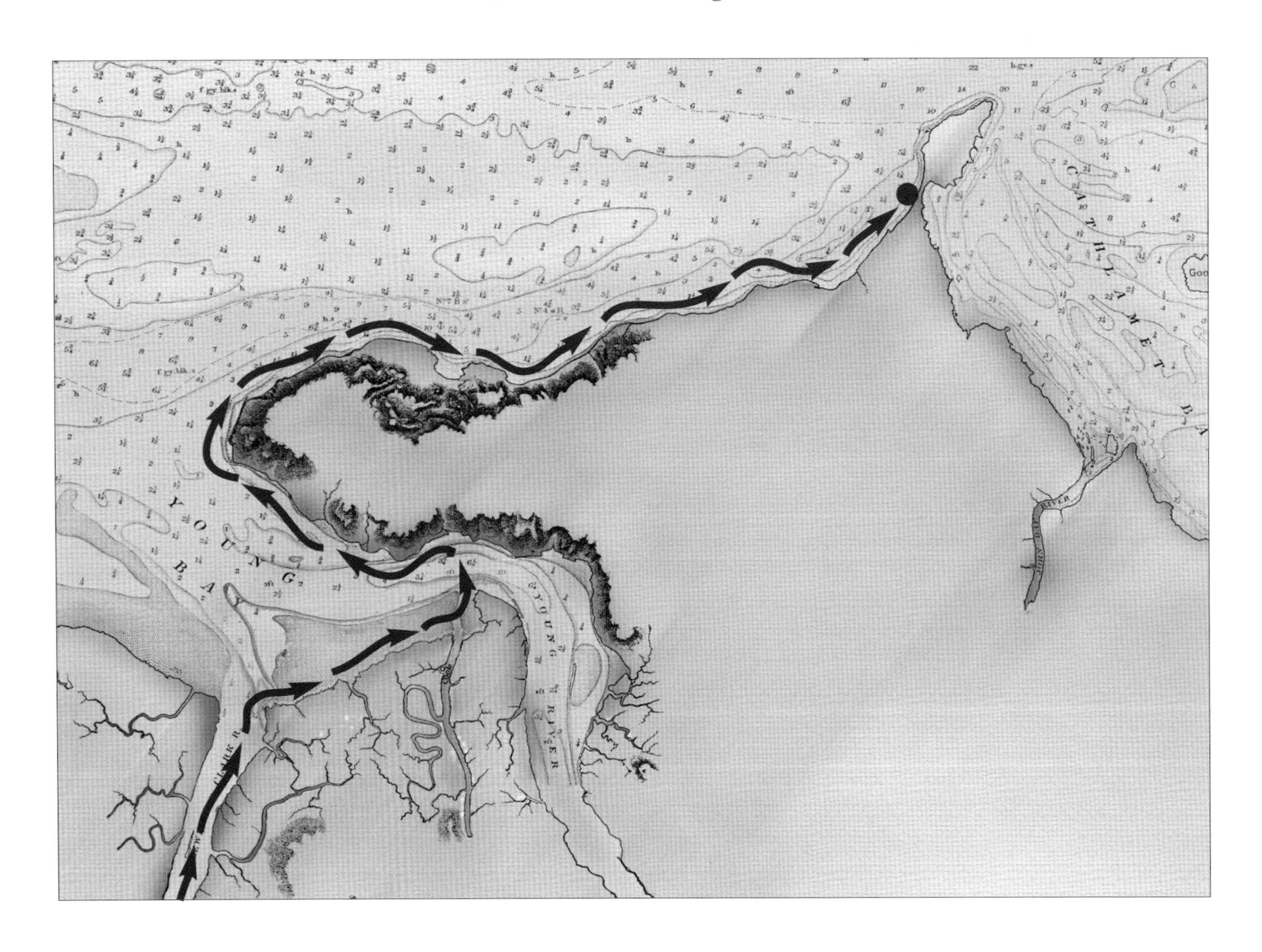

Lewis and three of his party return to Clark's camp in the afternoon.

After this was done, Lewis next had to find a very special piece of land where they could construct their winter camp. An ideal site would possess the following five features:

1. Since everything was transported by canoes, the campsite had to be close to a river;

2. At the same time, it had to be on high land, well above flood water;

3. The site had to be near the ocean for easy access to saltwater;

4. It had to be situated in the center of good elk hunting country;

5. Finally, there had to be an abundant source of fresh water nearby, preferably a cold spring.

Unfortunately, what Lewis encountered as he explored the area around the bay were thousands of acres of low and swampy land. Much of it was crisscrossed with a maze of small creeks and impassable sloughs, with most high ridges more than a mile away from any river. The distance to the ocean was uncertain, and the water was brackish.

It is impossible to know how many miles they walked and paddled, how many times they landed their canoe to explore, and how many sites they rejected; but finally Lewis found the precise piece of land that had all the elements they needed. It was an amazing accomplishment, and it explains what kept him away for so many days.

Lewis described the campsite he had found with its many excellent features, and everyone immediately agreed to go there.

this was verry Satisfactory information to all the party. we accordingly deturmined to proceed on to the Situation which Capt. Lewis had Viewed as Soon as the wind and weather Should permit and Comence building huts Moulton, *Journals* 6:108-109 (Clark)

The men were ready to move immediately, but the weather and the river refused to cooperate with their plans.

Rain continued . . . accompanied with hard wind from the SW. which provents our moveing from this Camp Moulton, *Journals* 6:108 (Clark)

Perhaps tomorrow the waters would flatten enough for them to slip downriver and into the site Lewis had found. The men wrapped up against the cold, drenching rain and bedded down for another long, weary night.

Despite this discomfort, Lewis and Clark's party must have felt relieved. After months of travel, capped by this unexpectedly grueling ordeal near the ocean, the men knew that they would soon be enjoying regular meals with a dry place to sleep. All they had to do was move downriver a couple of miles and build a winter camp on the site Captain Lewis had found.

Clark was the most relieved of them all. Lewis's long absence had frightened him. The two men relied heavily on each other's opinions, knowledge, and experience, so the thought of suddenly losing this partnership and having to lead the expedition back alone must have secretly worried him.

CHAPTER TEN

A Winter Camp at Last

the most eligible Situation for our purposes

Friday, December 6th

A small gale-force storm hits the coast. Rain and hard wind dominate the weather.

Daytime Low Tide:	*6:33 am*	*3.2*
Daytime High Tide:	*12:32 pm*	*9.9*
Sunrise:	*7:43am*	
Sunset:	*4:30 pm*	

MUCH TO LEWIS AND CLARK'S disappointment, another Pacific storm came ashore during the night. The wind howled, and thick gray clouds poured buckets of rain until dawn. The Columbia was stirred up into a pitching, rolling frenzy. A single glance told the men that they would not be going anywhere that day.

> *The wind blew hard all the last night with a moderate rain, the waves verry high*
>
> Moulton, *Journals* 6:109 (Clark)

By now the men were becoming familiar with the weather patterns. They had learned it was useless even to try to move during such a storm. All they could do was wait, and this storm, like the others, would pass.

> *the wind which is Still from the S W increased and rained Continued all day* Moulton, *Journals* 6:109 (Clark)

Though Lewis and Clark were learning the pattern of these Pacific Northwest storms, the rhythm of the tides continued to elude them. The tides moved up and down, but some were higher than others, and others much lower than the rest. It was confusing and sometimes caught them off guard. On this day, because of the full moon, the tide rose unexpectedly high, overflowed into their camp, and forced them to move to higher ground.[1]

> *High water to day at 12 oClock & 13 Inches higher than yesterday. we were obliged to move our Camp out of the Water on high grown all wet*
>
> Moulton, *Journals* 6:109 (Clark)

If the weather followed its typical pattern, tonight the storm would pass, and tomorrow they could finally move on downriver.

The waves prevent the party from moving downriver.

Saturday, December 7th

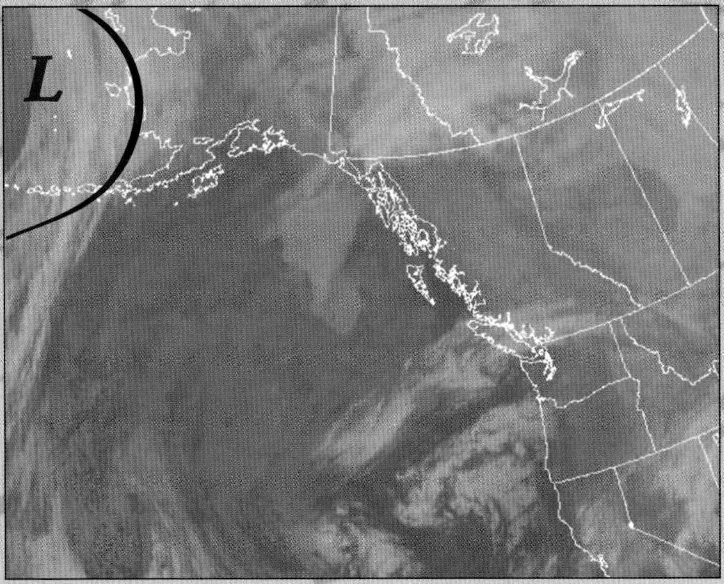

Rain continues throughout most of the night, fair at daylight. Wind diminishes. Another low pressure system builds in the North.

Daytime Low Tide:	*7:25 am*	*3.2*
Daytime High Tide:	*1:19 pm*	*10.0*
Sunrise:	*7:44 am*	
Sunset:	*4:29 pm*	

The small storm blew over during the night. The river remained stirred up, but Lewis and Clark knew they had to take this opportunity before another storm hit. The canoes were loaded and immediately launched.

> *this morning fair, have every thing put on board the Canoes and Set out to the place Capt Lewis had viewed . . . well Situated for winter quarters*
>
> Moulton, *Journals* 6:114 (Clark)

This would not be an easy journey. High waves rolled and splashed as their heavily loaded canoes lumbered through the swells, plunging dangerously deep into the cold, salty water. The captains were taking a tremendous risk leading their party downriver through such waters.

> *we proceeded on against the tide . . . the waves verry verry high, as much as our Canoes Could bear*
>
> Moulton, *Journals* 6:114 (Clark)

Along the way they were surprised to encounter Sergeant Pryor and his party walking upriver towards their camp. This was a very fortunate encounter, because the two parties could easily have missed each other. The waves were too high for the canoes to approach the shoreline, so they shouted orders to meet them downriver, somewhere along the bay.

> *we proceeded on around the point into the bay and landed to take brackfast . . . here all the party of Serjt Pryors joined us*
>
> Moulton, *Journals* 6:114 (Clark)

Pryor related his sad story. Instead of packing the elk to the bay, as Clark had instructed him, he had spent most of his time stumbling around the immense dark rainforest, once again completely lost and confused.

> *Sergt Pryor informed us that he had found the Elk, which was much further from the bay than he expected, that they missed the way for one day and a half, & when he found the Elk they were mostly Spoiled*
>
> Moulton, *Journals* 6:114 (Clark)

Even worse, in addition to becoming lost himself, Pryor had somehow managed to lose Clark's servant York.

> *my man york, who had Stoped to rite his load and missed his way*
>
> Moulton, *Journals* 6:114 (Clark)

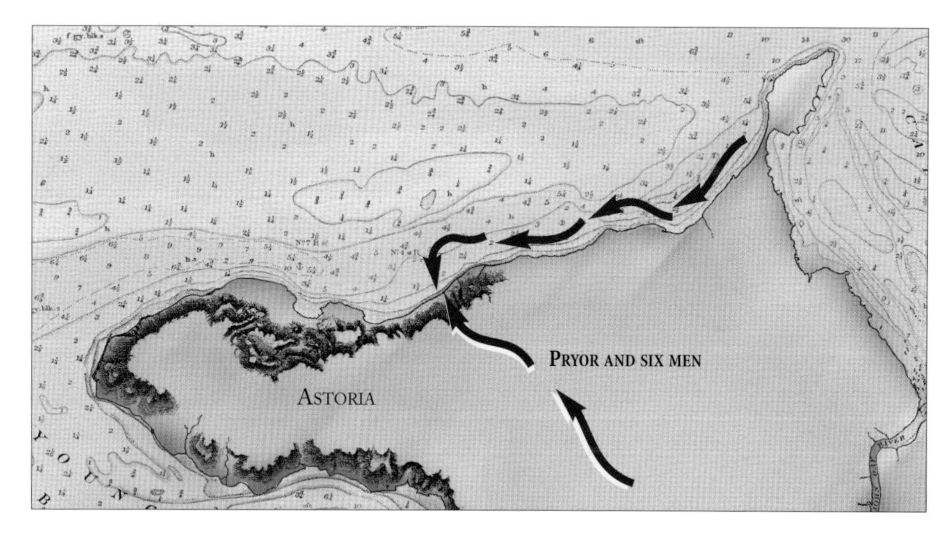

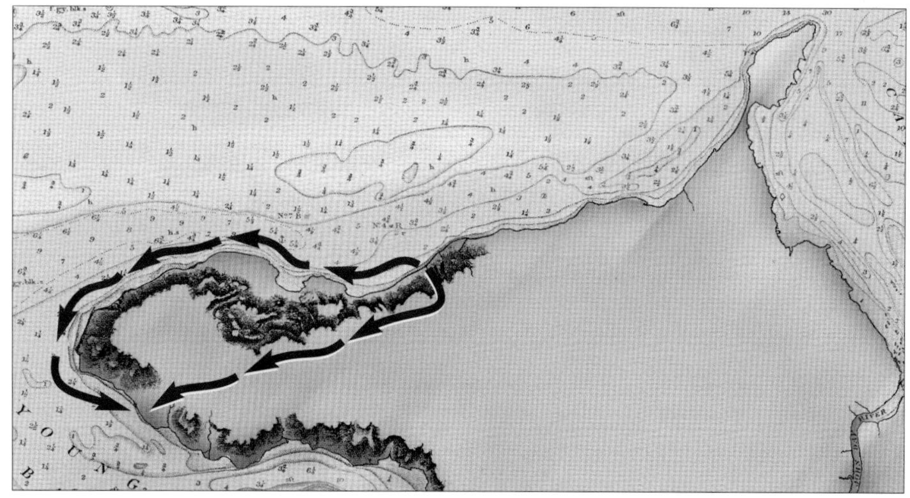

(top) All five canoes are loaded and the party sets out. Along the way they encounter Sgt. Pryor and his party. High waves prevent Lewis and Clark from landing so they agree to meet him around the point.

(bottom) Sgt. Pryor finds the party waiting for him.

After York rejoins the party, they paddle around the bay and enter a river that leads to the campsite Lewis found.

York could not have been far behind. Clark's men probably fired their rifles repeatedly into the air to signal York, then whooped and hollered into the woods,[2] because York soon found his way to the canoes and rejoined the party.

> *after brackfast I delayed about half an hour before York Came up, and then proceeded around this Bay*
>
> Moulton, *Journals* 6:114 (Clark)

Now they were all together once again. The men loaded their canoes and set off. This was, without any question, the last time they would load up these canoes this year.

They paddled around the lip of the bay, turned, and entered a wide tidewater river that led towards the distant hills. Everyone must have been wide-eyed, closely observing every detail of the surrounding landscape. This area was half swamp and half rainforest; every inch of ground was buried beneath thick, green plants and towering trees. Unless they learned this terrain intimately, they might spend hours, maybe even days, walking around in circles, completely lost, just like Sergeant Pryor.

> *we assended a river . . . 3 miles to the first point of high land on the West Side, the place Capt. Lewis had viewed and formed in a thick groth of pine about 200 yards from the river, this situation is on a rise about 30 feet higher than the high tides leavel and thickly Covered with lofty pine*
>
> Moulton, *Journals* 6:114 (Clark)

The two hunters whom Lewis had left behind greeted their companions as they approached. These men told the party exactly what they were hoping to hear: elk were everywhere.[3]

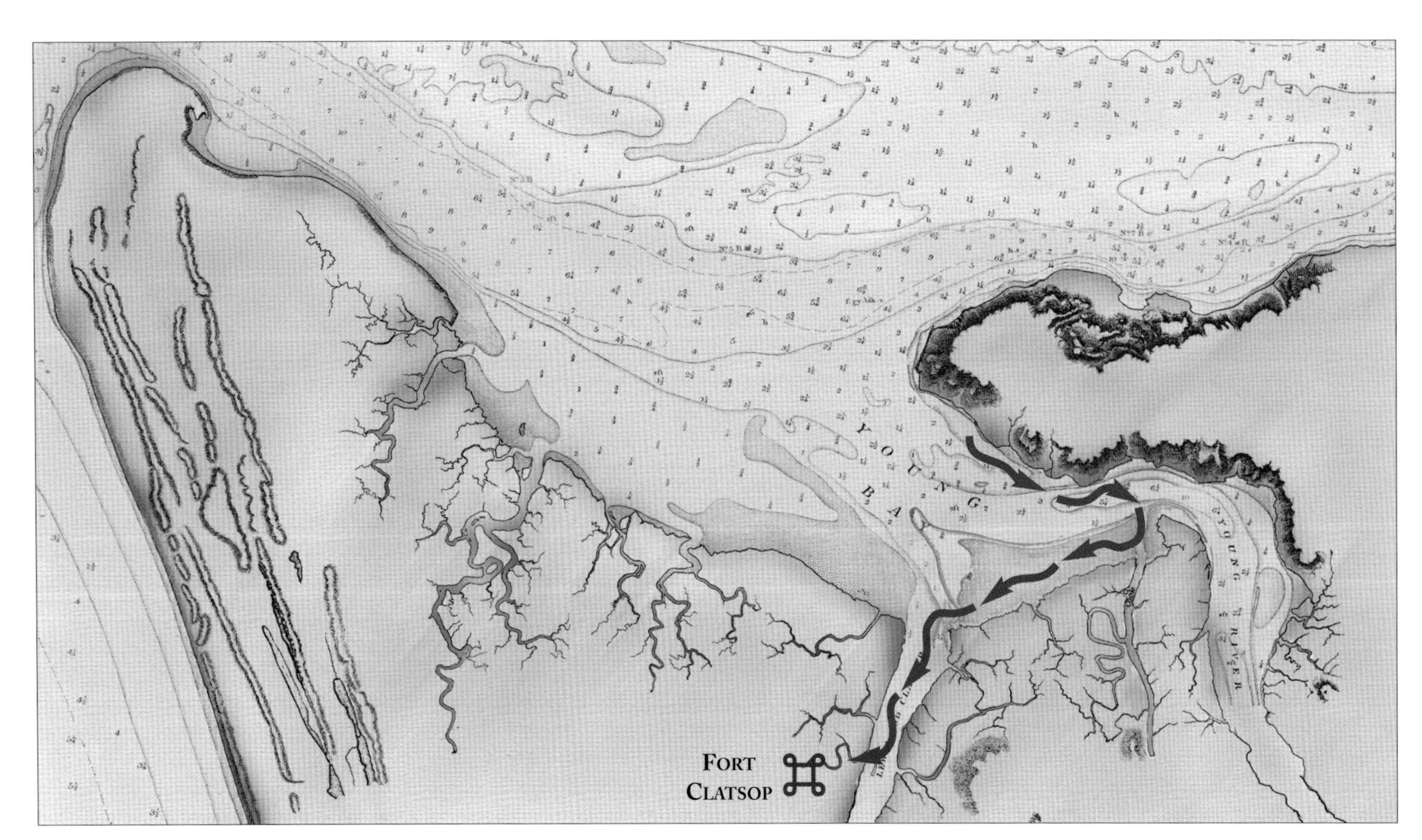

The situation is in the Center of as we conceve a hunting Countrey Moulton, *Journals* 6:112 (Clark)

Lewis had done a remarkable job of locating this particular winter camp. The site was thirty feet above any flood, yet only two hundred yards from the river; elk were all around. The thick rainforest provided plenty of timber for the construction of shelters and firewood, and a couple of yards away sweet fresh water bubbled out of the ground. Clark was immediately convinced that this would be an ideal location for a winter camp.

this is certainly the most eligable Situation for our purposes of any in its neighbourhood

Moulton, *Journals* 6:114 (Clark)

The canoes were unloaded and all the baggage carried up to the site. Lean-to sheds were hastily constructed to shelter the men from the incessant rain. Axes were sharpened and the ground cleared of brush, fallen limbs, and rotten trees. Clark drew out a floor plan of the building he wanted the men to construct. It was a perfect square, fifty feet by fifty feet, with seven rooms in the interior. This would be their home for the next several months, which they would name Fort Clatsop, after the local Indian tribe.

As each tree fell, more daylight flooded onto the dark, shadowy forest floor. This was the first time this piece of earth had been illuminated. Brilliant green mosses, delicate lichens, and lacy ferns grew everywhere, but the men had no time to admire this splendid beauty. Winter was here. There was no time to delay. They had to work at breakneck speed to build a dry shelter from the damp Pacific Northwest winter.

Winter is here; rain falls practically every day along the Pacific Northwest coast.

Lewis and Clark's party leave Fort Clatsop on March 23. They paddle upriver 140 miles against the powerful current of the Columbia, but lose a canoe in a mishap. Rather than continue to struggle against the river, the captains decide to trade their canoes for horses and follow the river.

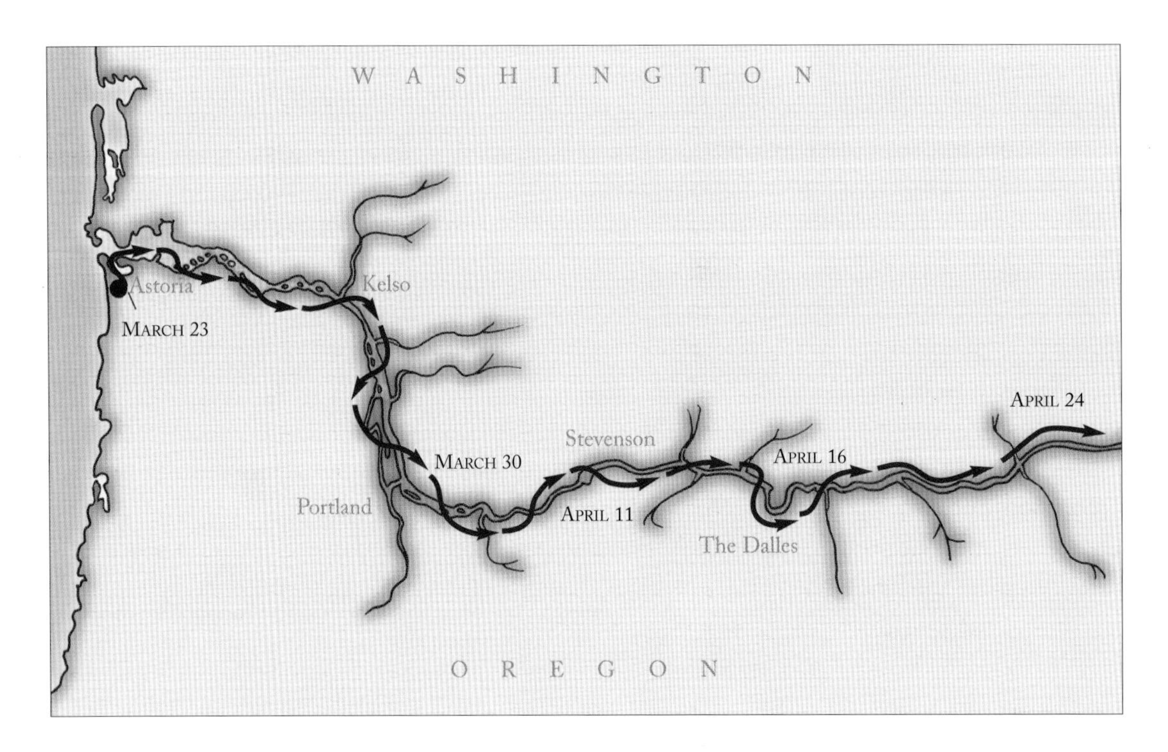

Their journey along the northern shore is hot, dry, and dusty. Food is scarce. Following an Indian route, the party crosses the Columbia and takes a short-cut overland, across the Snake River and up the Clearwater River to a wide valley floor. They camp here until the high mountain passes are clear of snow.

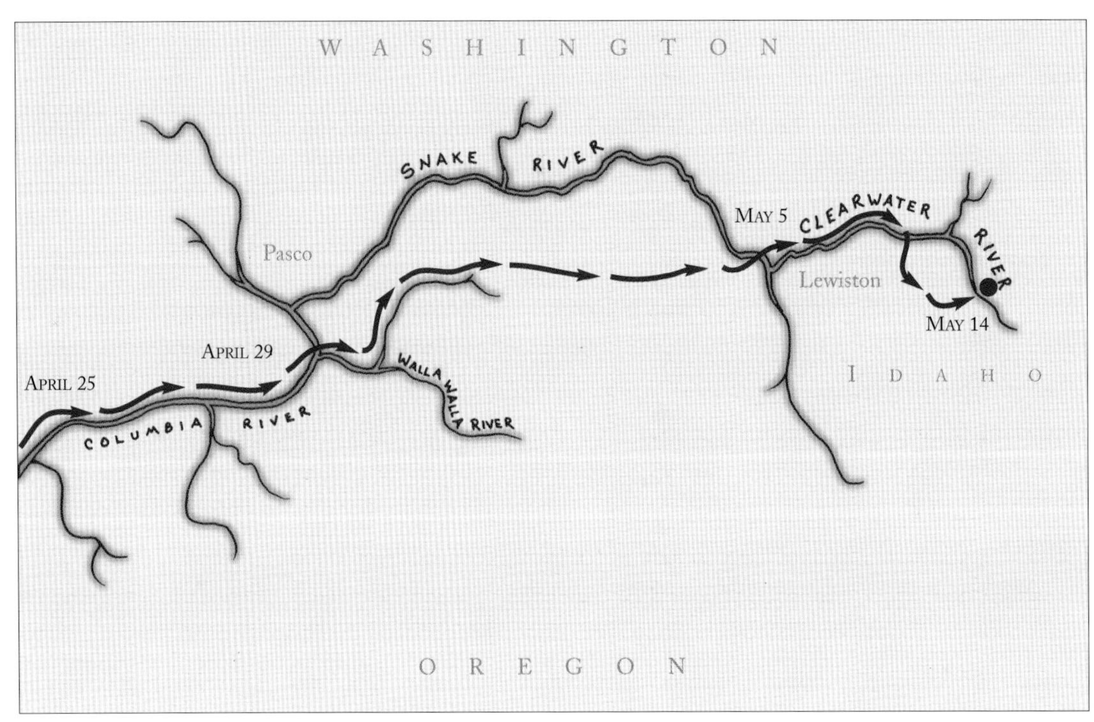

EPILOGUE

THE CONSTRUCTION OF LEWIS AND CLARK'S winter camp occurred at an unbelievably rapid pace. Two weeks after the first tree was cut, the captains' quarters were finished, complete with walls, roof, and floor. Several days later, the remaining members of the party moved into their rooms. By Christmas, Fort Clatsop had become a dry, secure place for them to pass the winter.

Hunters were sent out practically every day into the surrounding woods in pursuit of elk. This essential animal provided the party not only with food but also with hides that were tanned and sewn into clothing. Whale blubber, roots, dried berries, and fish provided an occasional break in this repetitious diet of elk meat, which was seasoned with salt from the ocean.

Throughout this idle winter, Lewis wrote extensively in his journal. As requested by President Jefferson, he recorded the weather on a daily basis, documented the local natives, and described the flora and fauna of the West. Both Lewis and Clark drew pictures of Indian canoes, condors, fish, and rare plants to help Jefferson understand what they had seen.

Their planned departure date was April, but elk became scarce and the men were falling ill, so they set out on March 23. The party struggled upriver with their canoes against the flow of the powerful Columbia for nearly two hundred miles, then switched to horses, preferring an overland route.

They reached the foothills of the Bitterroot Mountains by May, but deep snows delayed their crossing until June. Safe at last on the eastern side of the mountains, the captains decided to explore a much broader region during their return. They split up the party: Lewis with half the men headed north, Clark and the others went south towards the Yellowstone River. Five weeks later and five hundred miles from where they last were together, the two parties rejoined once again.

Charbonneau, Sacagawea and their baby left the party when they reached the friendly Mandan village; John Colter decided to go back upriver trapping. This reduced the members of the party by four. However, chief Sheheke accepted an invitation to visit President Jefferson as long as his family and interpreter could accompany him; his entourage added seven new people to the party.

They continued down the Missouri River, the flow of the river pushing them rapidly towards home. On September 23, exactly six months after leaving Fort Clatsop, the Lewis and Clark party arrived in St. Louis. Meriwether Lewis immediately wrote the president a letter.

Sir,

It is with pleasure that I anounce to you the safe arrival of myself and party at 12 OClk. today at this place with our papers and baggage. In obedience to your orders we have penitrated the Continent of North America to the Pacific Ocean [1]

St. Louis, September 23rd, 1806.

The Corps of Discovery

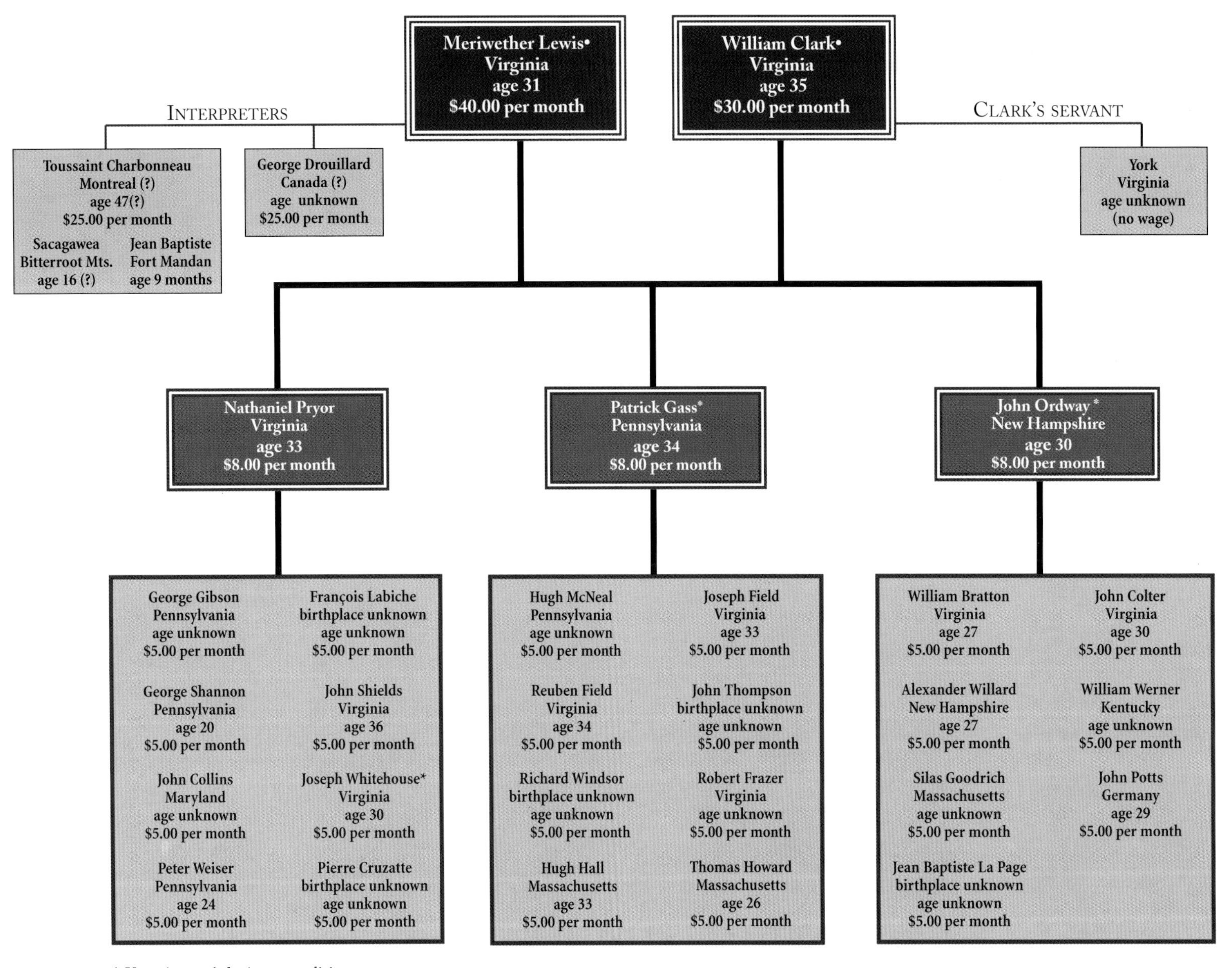

* Kept journal during expedition.

Appendices

To See or Not to See

Fur Trade around the World

Ship Repair

Jefferson's Letter to Lewis, 1803

Jefferson's Letter of Credit

Jefferson's Vocabulary List

The Tragic Loss

Elk or No Elk

York and Sacagawea

The Chinook

APPENDIX ONE

TO SEE OR NOT TO SEE

ON NOVEMBER 7, 1805, LEWIS AND CLARK'S party reached a significant milestone. They beached their canoes along the shore of the lower Columbia, looked downriver, and saw their destination. "Great Joy in Camp, Ocien in View," Clark wrote in his journal.

However, historians have repeatedly visited the site during the past century, and they all agree that Clark made a blunder. The ocean cannot be seen from this campsite.

The best way to understand the historians' conclusions is to look at a map. If we superimpose Clark's line of sight over a chart of the lower Columbia River, it is obvious that Clark could not have seen the ocean.

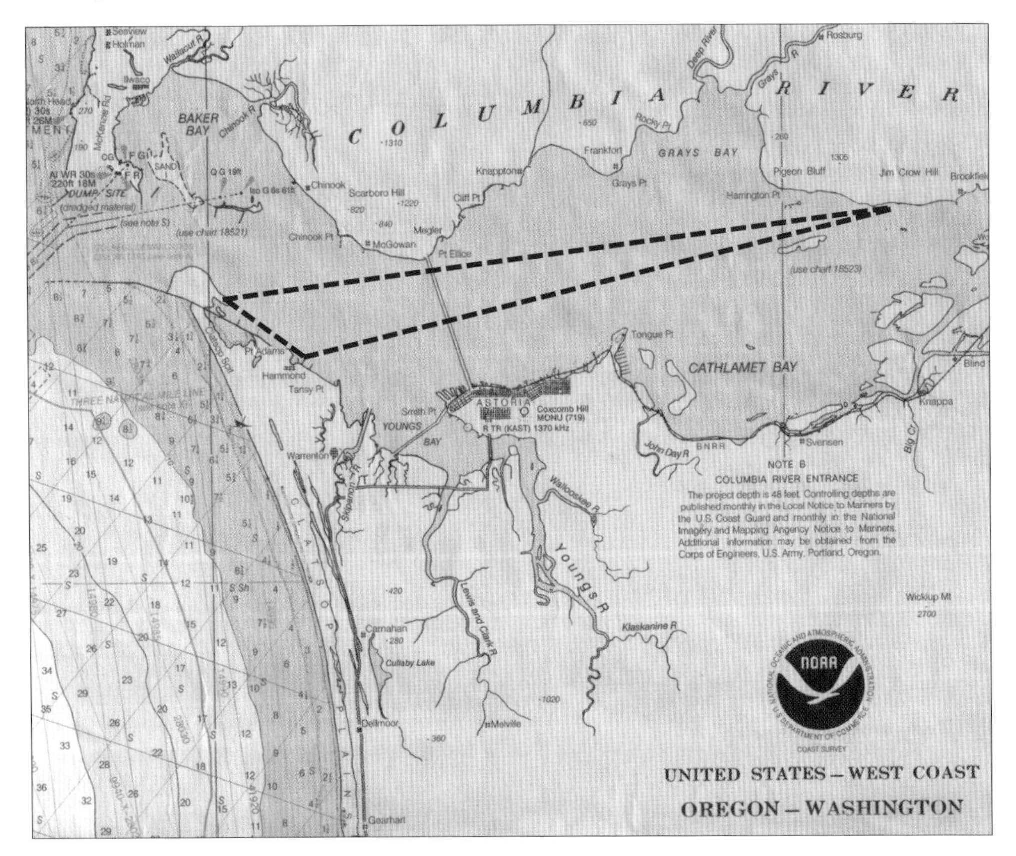

This contemporary NOAA map clearly shows that the ocean is not visible from Lewis and Clark's campsite of November 7, 1805.

However, it does seem odd that Lewis and Clark would make such a colossal mistake. These two men are constantly praised for their astute observations and remarkable discoveries. How could they have mistakenly identified the world's largest ocean?

Adding to the mystery is the conflict among the leading Lewis and Clark scholars regarding which body of water the party did see from their campsite. Some claim that they were looking at Grays Bay, which, however, happens to be around a point and is therefore impossible to see without a helicopter. Others assume that Lewis and Clark mistook the estuary for the ocean, which makes even less sense, especially since they could see the hills of Astoria to the west of it.

This confusing problem piqued my curiosity, so I decided to undertake my own investigation. After several years of studying all the explanations and theories, I found an answer that will no doubt surprise many historians. I believe that Lewis and Clark did see the ocean, precisely as they claimed in their journals, on November 7, 1805. They did not make a mistake.

The crucial piece of information missing from previous discussions of this problem, and the one on which I have based my conclusion, is that the mouth of the Columbia River is not now in the same position as it was in 1805. The mouth of this great river has moved.

Moving the Mouth of the Columbia

As settlers flooded into the Oregon Territory in the 1850s, the Columbia River became increasingly important for ship traffic. However, the river's treacherous mouth was a major obstacle. Ships could not safely enter with supplies nor leave with produce; disasters became commonplace and slowed the region's growth. After years of study, surveys, and debate, the government decided to remedy this problem by constructing an enormous breakwater.

A railroad trestle was built out into the ocean for a distance of over four miles. It resembled a bridge that dead ended beyond the surf. Simultaneously, farther upriver, mountains of rock were dynamited, loaded on barges and floated towards Point Adams. This rock was reloaded onto trains that ran out into the ocean.

Between 1885 and 1895, millions of tons of rock were placed in the ocean. Eventually the rock settled and its level rose above the waves. The construction of this "South Jetty" was the largest public works project in the history of the United States.

The jetty changed the flow of the Columbia River, deepening the channel for ships and blocking the surge of ocean waves. Sand immediately

A fifteen-ton boulder is loaded onto a dumping flatcar.

Two locomotives and twenty flatcars loaded with boulders head out into the ocean.

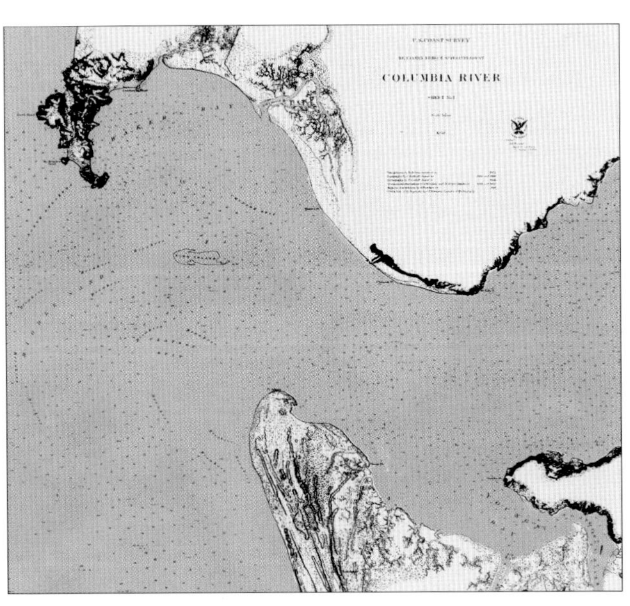

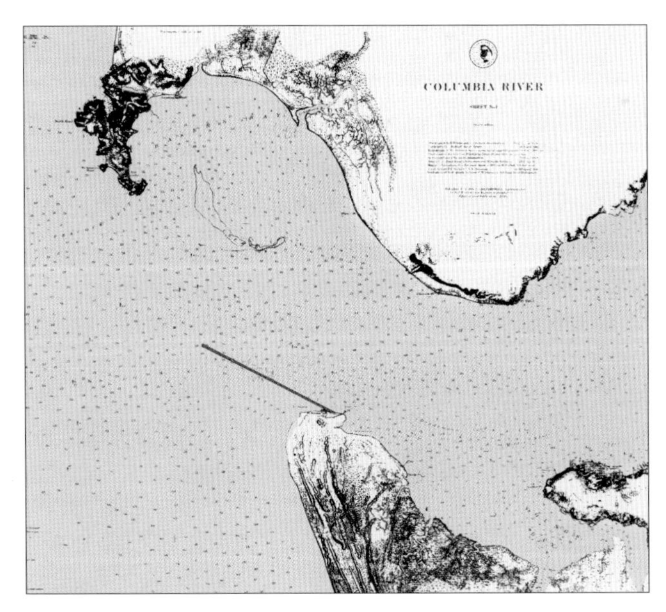

U.S. Coast Survey maps: Mouth of the Columbia River.

(left) 1870: Mouth of the Columbia River is unchanged since the time of Lewis and Clark's arrival there.

(right) 1890: The construction of the South Jetty was underway.

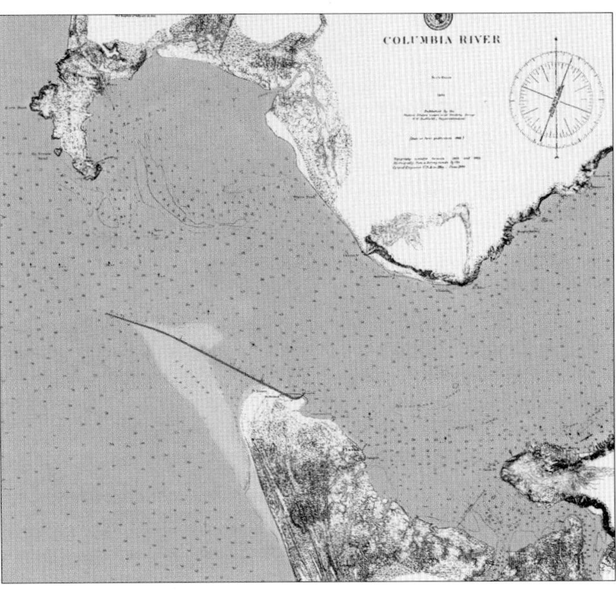

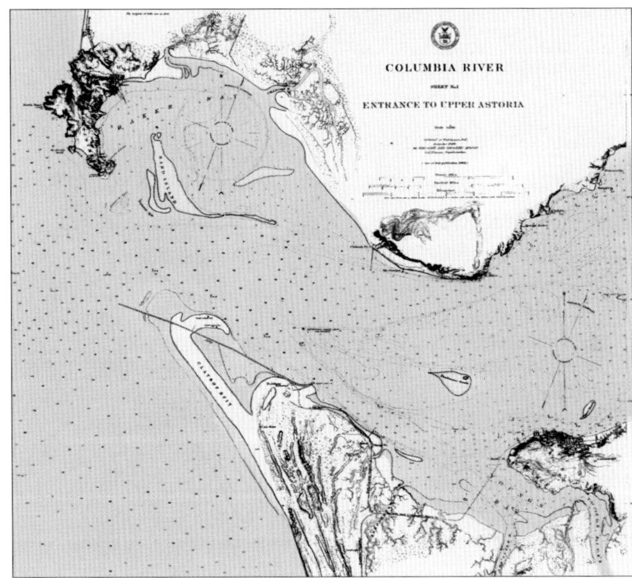

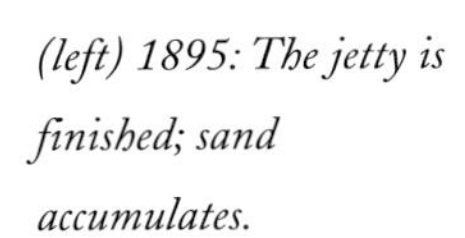

(left) 1895: The jetty is finished; sand accumulates.

(right) 1905: Ten years after construction was finished, Point Adams is no longer the northwest corner of Oregon.

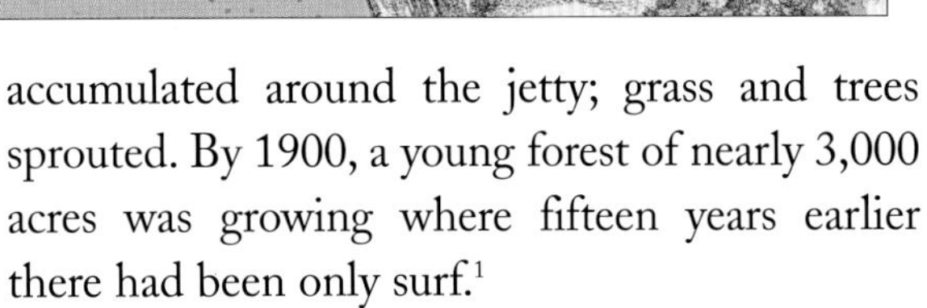

accumulated around the jetty; grass and trees sprouted. By 1900, a young forest of nearly 3,000 acres was growing where fifteen years earlier there had been only surf.[1]

Years after the mouth of the Columbia had been altered, preparations for the Lewis and Clark centennial began. Writers retraced the route of the expedition, and the first edition of Lewis and Clark's complete journals was prepared for publication. The editor, Reuben Gold Thwaites, relied on secondhand information when he included a little footnote beneath the date of November 7 which read:

> *The ocean could not possibly be seen from this point* [2]

From that moment on, generations of historians have read this footnote, assumed that it was accurate, and repeated it in their own books; and, of course, any scholar who came to see for

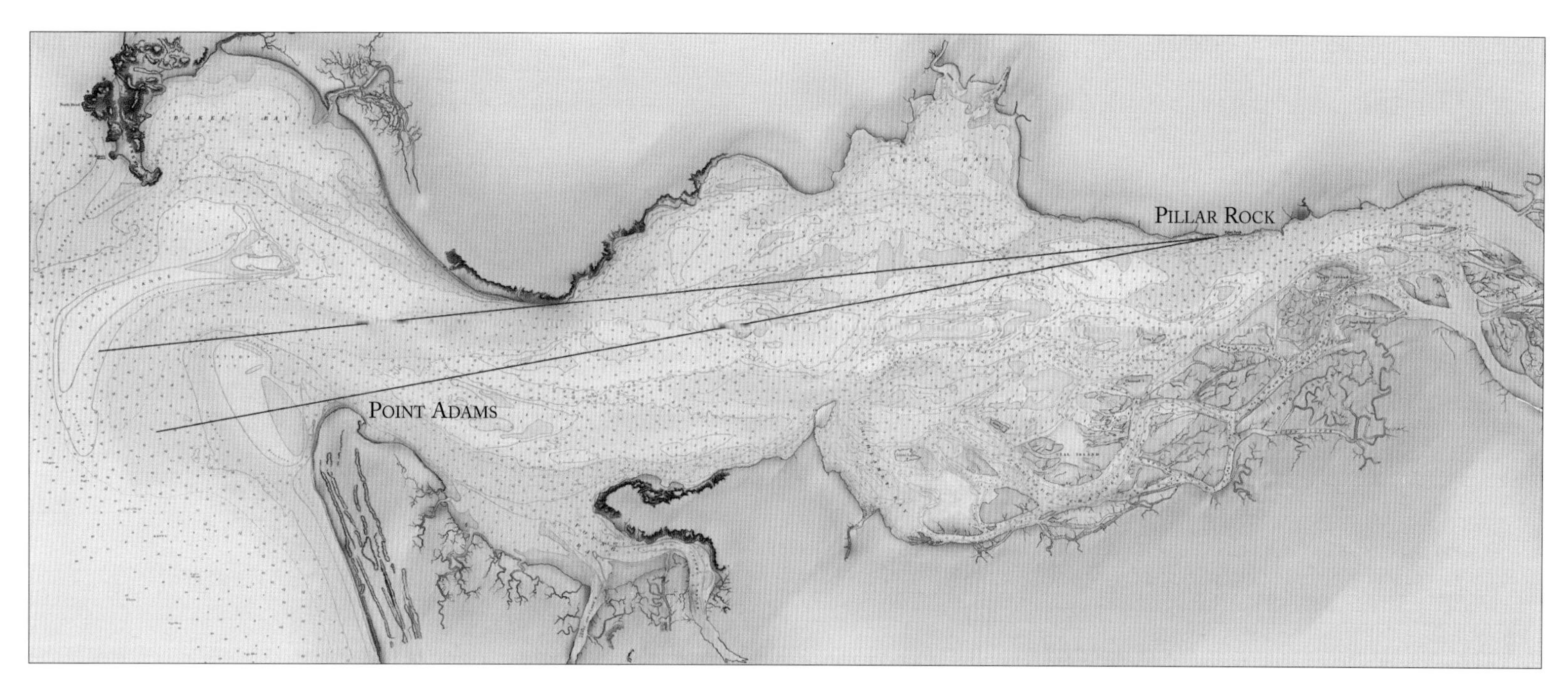

A map of the Columbia River drawn in 1876, with Clark's line of sight superimposed, clearly shows that their campsite of November 7, 1805 was perfectly aligned with the mouth of the Columbia River.

himself would find that from the campsite the ocean is certainly not in view.

However, anyone examining an accurate chart drawn before the construction of the South Jetty, with a line of sight superimposed over it, would see that Lewis and Clark did, in fact, have a direct view of the ocean.

Further evidence that Lewis and Clark saw the ocean on November 7 can be found in their journals.

Additional Evidence

After leaving their camp of November 7, the party continued down to the coast and then, eighteen days later, returned to this same campsite. At that point there was no mystery about the landscape downriver; Lewis and Clark knew exactly what they were seeing as they looked towards the ocean. Obviously, if they felt that they had made an error, this would have been a perfect time to correct it, but a close examination of the journals reveals no corrections.

Six days later, while forced to halt along the shoreline, Clark described the sound of the crashing surf and claimed they had heard this noise ever since they had first come into view of the ocean on November 7. He writes:

> *roars like a repeeted roling thunder and have rored in that way ever Since our arrival in its borders which is now 24 Days Since we arrived in Sight of the Great Western Ocian* [3]

To Clark, "24 days" is not an arbitrary number. He had to do some calculating to arrive at that number, counting backwards from December 1 and arriving precisely at November 7. This confirms that, even with 20-20 hindsight, after having been downriver to the ocean and back, Clark still agreed with his original observation. They first saw the ocean on November 7.

APPENDIX TWO

FUR TRADE AROUND THE WORLD

"The Golden Round" Fur traders sailed out of Boston, around South America, then up to the Pacific Northwest. Loaded up with furs, they crossed the ocean and sold their cargo in China. The profits were used to purchase silk, tea, spices, and porcelain. These exotic, highly-coveted goods were then carried around the tip of Africa and back into Boston harbor.

IN THE 1770S, ON CAPTAIN COOK'S THIRD and final voyage, his crews accidentally discovered that Chinese merchants would pay a fortune for sea otter fur from the Northwest Coast of America.[1] When word of this reached England and America, ships were sent out to investigate. It was soon discovered to be true, and by the early 1790s the Northwest Coast was experiencing its first gold rush.[2]

Ships, loaded with tons of trade goods, would leave Boston in the fall. They crossed to the Canary Islands where strong winds pushed them southward along the coast of Brazil, and then on to the Falkland Islands. From there they rounded the tip of South America and immediately headed northward into the Pacific Ocean. Their first stop was Hawaii, where they replenished their supply of meats, fruit, and potatoes. It was now the month of February or March and the crews were six months into their voyage.

From Hawaii the ships sailed towards the Northwest Coast. They reached the Columbia River, Vancouver Island, Queen Charlotte Islands, or Nootka Sound around April, then cruised from cove to cove in search of natives who possessed sea otter furs. The greatest abundance

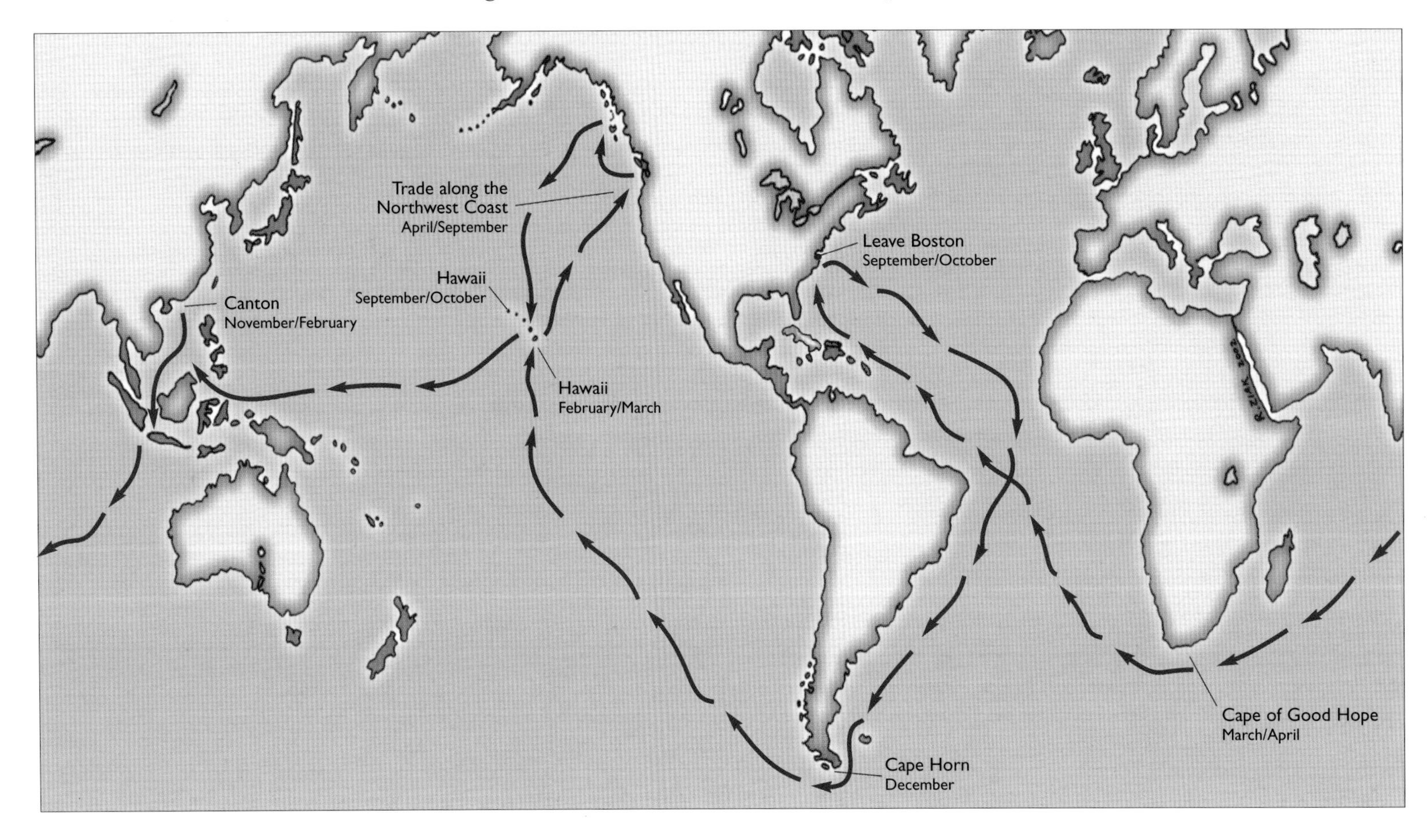

of these creatures was found in northern waters, but many tribes acquired the furs in anticipation of trade with the "Boston Men."

The ships concluded their trading before the stormy winter weather arrived along the Northwest Coast. The sailors returned to Hawaii, replenished their supplies, then sailed directly to Canton, China. The furs were sold, and with the profits the captains purchased tea, porcelain, silk, and fine cotton cloth. Sixteen months had now elapsed since they had left Boston, and they were still thousands of miles away from home.

By February it was time to go home. They sailed from Canton, passed Sunda Straight, crossed the Indian Ocean, and rounded the Cape of Good Hope by late spring. From the tip of Africa the route back to Boston was easy.

Some ships did not return; the vessel and crew were lost at sea. However, for those who survived, a 200 percent profit was not unusual, and many Boston investors earned from 300 to 500 percent. Records exist that indicate that some lucky captains, who purchased furs at extremely low prices, brought their owners a profit of 2000 percent.[3]

Fur traders sailed up and down the coastline searching for protected bays and inlets inhabited by native people. When the Indians noticed a ship at anchor, they would load furs into their canoes and paddle out to begin trading.

APPENDIX THREE
SHIP REPAIR

IN ORDER TO ARRIVE ALONG THE NORTHWEST Coast where valuable furs could be acquired, the sailing ships had to brave thousands of miles of ocean. It was a risky voyage. Sometimes the storms they encountered severely damaged their ships. For example, a clerk named John Hoskins described the damage their ship *Columbia* received on their way to the Northwest Coast when they encountered a storm off the coast of Brazil.

> *in a heavy squall or tornado, into the westward; which struck the Ship, carried away the fore and maintopgallantmasts, the foretopsail yard, cross and trussel trees; sprung the mizentopmast; damaged the maintopgallant sail, main topsail yard, cross trees, etca. etca.*[1]

If these crippled ships were able to reach the Northwest Coast, the crew began repairs as soon as the ship came to anchor. Tall slender trees would be chopped down and hewn into replacement parts. While exploring around Cape Disappointment Lewis possibly noticed that timber had been felled by ship's carpenters. He wrote:

> *The bay in which this trade is carryed on is spacious and comodious . . . fresh water and wood are very convenient and excellent timber for refiting and reparing vessels.*[2]

Before crossing from Nootka Sound to China the crew of the *Columbia* made many spare parts for their ship in 1792. Some chopped firewood and filled casks with drinking water, while others prepared new rigging from the coastal forest.

> *Sent a strong party on shore wooding and cutting spars. Took off a boat load of wood.*
>
> *A party wooding. Cut a maintopmast, a topsail yard, two topgallant masts, topgallant yards, boat masts, sprits, booms, boat hooks, etc."*
>
> *On the 2d we filled our water, completed our wood, and cut a number of logs for plank.*[3]

The forests around Cape Disappointment were excellent for ship repair. The heavy rainfall combined with thin topsoil forced the trees to grow slowly, creating remarkably strong wood; the competition for sunlight beneath the heavy forest canopy encouraged them to grow as straight as flagpoles. It is very likely that the ship's carpenters would have remarked that the trees growing along this coastline were the most perfect lumber they had ever seen.

The masts and yardarms of sailing ships must endure tremendous force when set with full sail. Despite the fact that they were designed for this purpose, they frequently needed repair. In fact, every ship carried its own carpenter, who was kept busy constantly repairing the vessel from top to bottom.

APPENDIX FOUR

JEFFERSON'S LETTER TO LEWIS, 1803

TO MERIWETHER LEWIS, ESQUIRE, CAPTAIN OF THE 1ST REGIMENT OF INFANTRY OF THE UNITED STATES OF AMERICA: Your situation as Secretary of the President of the U.S. has made you acquainted with the objects of my confidential message of Jan. 18, 1803 to the legislature; you have seen the act they passed, which, tho' expressed in general terms, was meant to sanction those objects, and you are appointed to carry them into execution.

Appointed as leader

Supplied with materials and men

Instruments for ascertaining, by celestial observations, the geography of the country through which you will pass, have been already provided. Light articles for barter and presents among the Indians, arms for your attendants, say for from 10. to 12. men, boats, tents, & other traveling apparatus, with ammunition, medecine, surgical instruments and provisions you will have prepared with such aids as the Secretary at War can yield in his department; & from him also you will recieve authority to engage among our troops, by voluntary agreement, the number of attendants above mentioned, over whom you, as their commanding officer, are invested with all the powers the laws give in such a case.

As your movements while within the limits of the U.S. will be better directed by occasional communications, adapted to circumstances as they arise, they will not be noticed here. What follows will respect your proceedings after your departure from the United states.

Passage assured across foreign territories

Your mission has been communicated to the ministers here from France, Spain & Great Britain, and through them to their governments; & such assurances given them as to it's objects, as we trust will satisfy them. The country [of Louisiana] having been ceded by Spain to France, [and possession by this time probably given,] the passport you have from the minister of France, the representative of the present sovereign of the country, will be a protection with all it's subjects; & that from the minister of England will entitle you to the friendly aid of any traders of that allegiance with whom you may happen to meet.

Object of the mission

The object of your mission is to explore the Missouri river, & such principal stream of it, as, by it's course and communication with the waters of the Pacific ocean, whether the Columbia, Oregan, Colorado or any other river may offer the most direct & practicable water communication across this continent for the purposes of commerce.

Measure latitude and longitude

Beginning at the mouth of the Missouri, you will take observations of latitude & longitude, at all remarkeable points on the river, & especially at the mouths of rivers, at rapids, at islands, & other places & objects distinguished by such natural marks & characters of a durable kind, as that they may with certainty be recognised hereafter. The course of the river between these points of observation may be supplied by the compass the log-line & by time, corrected by the observations themselves. The variations of the compass too, in different places, should be noticed.

The interesting points of the portage between the heads on the Missouri, & of the water offering the best communication with the Pacific ocean, should also be fixed by observation, & the course of that water to the ocean, in the same manner as that of the Missouri.

Carefully record latitudes; make copies of notes

Your observations are to be taken with great pains & accuracy, to be entered distinctly & intelligibly for others as well as yourself, to comprehend all the elements necessary, with the aid of the usual tables, to fix the latitude and longitude of the places at which they were taken, and are to be rendered to the war-office, for the purpose of having the calculations made concurrently by proper persons within the U.S. Several copies of these as well as your other notes should be made at leisure times, & put into the care of the most trust-worthy of your attendants, to guard, by multiplying them, against the accidental losses to which they will be exposed. A further guard would be that one of these copies be on the paper of the birch, as less liable to injury from damp than common paper.

Gather information about native people

The commerce which may be carried on with the people inhabiting the line you will pursue, renders a knolege of those people important. You will therefore endeavor to make yourself acquainted, as far as a diligent pursuit of your journey shall admit, with the names of the nations & their number.

the extent & limits of their possessions;

their relations with other tribes of nations;

their language, traditions, monuments;

their ordinary occupations in agriculture, fishing, hunting,war, arts, & the implements for these;

their food, clothing, & domestic accomodations;

the diseases prevalent among them, & the remedies they use; moral & physical circumstances which distinguish them from the tribes we know;peculiarities in their laws, customs & dispositions; and articles of commerce they may need or furnish, & to what extent.

And, considering the interest which every nation has in extending & strengthening the authority of reason & justice among the people around them, it will be useful to acquire what knolege you can of the state of morality, religion, & information among them; as it may better enable those who may endeavor to civilize & instruct them, to adapt their measures to the existing notions & practices of those on whom they are to operate.

Collect wide-ranging scientific information

Others objects worthy of notice will be

the soil & face of the country, it's growth & vegetable productions, especially those not of the U.S.

the animals of the country generally, & especially those not known in the U.S.

the remains or accounts of any which may be deemed rare of extinct;

the mineral productions of every kind; but more particularly metals, limestone, pit coal, & saltpetre; salines & mineral waters, noting the temperature of the last, & such circumstances as may indicate their character;

volcanic appearances;

climate, as characterised by the thermometer, by the proportion of rainy, cloudy, & clear days, by lightning, hail, snow, ice, by the access & recess of frost, by the winds prevailing at different seasons, the dates at which particular plants put forth or lose their flower, or leaf, times of appearance of particular birds, reptiles or insects.

Gain knowledge of river systems

Altho' your route will be along the channel of the Missouri, yet you will endeavor to inform yourself, by enquiry, of the character & extent of the country watered by it's branches, & especially on it's Southern side. The North river or Rio Bravo which runs into the gulph of Mexico, and the North river, or Rio colorado which runs into the gulph of California, are understood to be the principal streams heading opposite to the waters of the Missouri, and running Southwardly. Whether the dividing grounds between the Missouri & them are mountains or flat lands, what are their distance from the Missouri, the character of the intermediate country, & the people inhabiting it, are worthy of particular enquiry. The Northern waters of the Missouri are less to be enquired after, because they have been ascertained to a considerable degree, & are still in a course of ascertainment by English traders, and travellers. But if you can learn any thing certain of the Northern source of the Missisipi, & of it's position relatively to the lake of the woods, it will be interesting to us.

Some account too of the path of the Canadian traders from the Missisipi, at the mouth of the Ouisconsing to where it strikes the Missouri, & of the soil and rivers in it's course, is desireable.

Treat the Indians well; offer to educate their children

In all your intercourse with the natives, treat them in the most friendly & conciliatory manner which their own conduct will admit; allay all jealousies as to the object of your journey, satisfy them of it's innocence, make them acquainted with the position, extent, character, peaceable & commercial dispositions of the U.S., of our wish to be neighborly, friendly, useful to them, & of our disposition to a commercial intercourse with them; confer with them on the points most convenient as mutual emporiums, and the articles of most desireable interchange for them & us. If a few of their influential chiefs, within practicable distance, wish to visit us, arrange such a visit with them, and furnish them with authority to call on our officers, on their entering the U.S. to have them conveyed to this place at the public expence. If any of them should wish to have some of their young people brought up with us, & taught such arts as may be useful to them, we will receive, instruct & take care of them. Such a mission, whether of influential chiefs or of young people, would give some security to your own party. Carry with you some matter of the kinepox; inform those of them with whom you may be, of it's efficacy as a preservative from the smallpox; & instruct & encourage them in the use of it. This may be especially done wherever you winter.

Bring the party home safely

As it is impossible for us to foresee in what manner you will be received by those people, whether with hospitality or hostility, so is it impossible to prescribe the exact degree of preserverance with which you are to pursue your journey. We value too much the lives of citizens to offer them to probable destruction. Your numbers will be sufficient to secure you against unauthorized opposition of individuals or of small parties: but if a superior force, authorised, or not authorised, by a nation, should be arrayed against your further passage, and inflexibly determined to arrest it, you must decline it's further pursuit, and return. In the loss of yourselves, we should lose also the information you will have acquired. By returning safely with that, you may enable us to renew the essay with better calculated means. To your own discretion therefore must be left the degree of danger you may risk, and the point at which you should decline, only saying we wish you to err on the side of your safety, and to bring back your party safe even if it be with less information.

Send letters when possible; encode sensitive information

As far up the Missouri as the white settlements extend, an intercourse will probably be found to exist between them & the Spanish posts of St. Louis opposite Cahokia, or Ste. Genevieve opposite Kaskaskia. From still further up the river, the traders may furnish a conveyance for letters. Beyond that, you may perhaps be able to engage Indians to bring letters for the government to Cahokia or Kaskaskia, on promising that they shall there receive such special compensation as you shall have stipulated with them. Avail yourself of these means to communicate to us, at seasonable intervals, a copy of your journal, notes & observations, of every kind, putting into cypher whatever might do injury if betrayed.

Question regarding shipment of furs

Should you reach the Pacific ocean inform yourself of the circumstances which may decide whether the furs of those parts may not be collected as advantageously at the head of the Missouri (convenient as is supposed to the waters of the Colorado & Oregan or Columbia) as at Nootka sound, or any other point of that coast; and that trade be consequently conducted through the Missouri & U.S. more beneficially than by the circumnavigation now practised.

Possible return by ship

On your arrival on that coast endeavor to learn if there be any port within your reach frequented by the sea-vessels of any nation, & to send two of your trusty people back by sea, in such way as shall appear practicable, with a copy of your notes: and should you be of opinion that the return of your party by the way they went will be eminently dangerous, then ship the whole, & return by sea, by the way either of cape Horn, or the cape of good Hope, as you shall be able. As you will be without money, clothes or provisions, you must endeavor to use the credit of the U.S. to obtain them, for which purpose open letters of credit shall be furnished you, authorising you to draw upon the Executive of the U.S. or any of it's officers, in any part of the world, on which draughts can be disposed of, & to apply with our recommendations to the Consuls, agents, merchants, or citizens of any nation with which we have intercourse, assuring them, in our name, that any aids they may furnish you, shall be honorably repaid, and on demand. Our consuls Thomas Hewes at Batavia in Java, Wm. Buchanan in the Isles of France & Bourbon & John Elmslie at the Cape of good Hope will be able to supply your necessities by draughts on us.

Letters of credit provided

Continue observations during return journey

Should you find it safe to return by the way you go, after sending two of your party round by sea, or with your whole party, if no conveyance by sea can be found, do so; making such observations on your return, as may serve to supply, correct of confirm those made on your outward journey.

Discharge party; return to Washington

On re-entering the U.S. and reaching a place of safety, discharge any of your attendants who may desire & deserve it, procuring for them immediate paiment of all arrears of pay & cloathing which may have incurred since their departure, and assure them that they shall be recommended to the liberality of the legislature for the grant of a souldier's portion of land each, as proposed in my message to Congress: & repair yourself with your papers to the seat of government.

If death is near, appoint a new leader

To provide, on the accident of your death, against anarchy, dispersion, & the consequent danger to your party, and total failure of the enterprize, you are hereby authorised, by any instrument signed & written in your own hand, to name the person among them who shall succeed to the command on your decease, and by like instruments to change the nomination from time to time as further experience of the characters accompanying you shall point out superior fitness: and all the powers and authorities given to yourself are, in the event of your death, transferred to, & vested in the successor so named, with further power to him, and his successors in like manner to name each his successor, who, on the death of his predecessor, shall be invested with all the powers & authorities given to yourself.

Given under my hand at the city of Washington this 20th day of June 1803.

Th Jefferson

APPENDIX FOUR

I Thomas Jefferson, President of the United States of America, have written this letter of general credit for you

THOMAS JEFFERSON REALIZED THAT LEWIS might need money if he reached the Pacific coast. Perhaps, if it were too dangerous to return by land, the party would board a ship back to America. On the other hand, maybe their supply of trade goods would need replenishing. In either case, carrying money was impractical, so Thomas Jefferson wrote this letter of credit for Lewis to use, if needed.

Washington. U.S. of America. July 4. 1803

Dear Sir In the journey which you are about to undertake for the discovery of the course and source of the Missouri, and of the most convenient water communication from thence to the Pacific ocean, your party being small, it is to be expected that you will encounter considerable dangers from the Indian inhabitants. Should you escape those dangers and reach the Pacific ocean, you may find it imprudent to hazard a return the same way, and be forced to seek a passage round by sea, in such vessels as you may find on the Western coast. But you will be without money, without clothes, & other necessaries; as a sufficient supply cannot be carried with you from hence. Your resource in that case can only be in the credit of the U.S. for which purpose I hereby authorise you to draw on the Secretaries of State, of the Treasury, of War & of the Navy of the U.S. according as you may find your draughts will be most negociable, for the purpose of obtaining money or necessaries for yourself & your men; and I solemnly pledge the faith of the United States that these draughts shall be paid punctually at the date they are made payable. I also ask of the Consuls, agents, merchants & citizens of any nation with which we have intercourse or amity to furnish you with those supplies which your necessities may call for, assuring them of honorable and prompt retribution. And our own Consuls in foreign parts where you may happen to be, are hereby instructed & required to be aiding & assisting you in whatsoever may be necessary for procuring your return back to the United States. And to give more entire satisfaction & confidence to those who may be disposed to aid you, I Thomas Jefferson, President of the United States of America, have written this letter of general credit for you with my own hand, and signed it with my name.[1]

APPENDIX SIX

JEFFERSON'S VOCABULARY LIST

THOMAS JEFFERSON COLLECTED INDIAN languages. In his spare time he analyzed and compared the vocabularies, looking for any words that might indicate a close relationship to languages spoken in Asia or Europe. It was part of a project he began in the 1770s.

Jefferson had a list of 324 words that he would send out into the wilderness with whoever had access to distant tribes. Presumably this would be trappers, merchants, explorers, and maybe government officials. Meriwether Lewis carried these lists for Jefferson and never missed an opportunity to have the words translated. While descending the Columbia Lewis sampled two languages on October 17 near the mouth of the Snake River. Clark writes:

> *Capt. Lewis took a vocabelary of the Language of those people who call themselves So kulk, and also one of the language of a nation . . . who Call themselves Chim na pum*[1]

Ten days later, near the Long Narrows, they sampled two more languages. Again Clark writes:

> *we took a vocabelary of the Languages of the 2 nations, the one living at the Falls call themselves E-nee-shur The other... call themselves E-chee-lute* [2]

> *the Languages of those two chiefs which are verry different notwithstanding they are Situated within Six miles of each other* [3]

On November 18 Lewis wrote down a Chinook vocabulary. Gass says:

> *The Indians still remained with us, and Capt. Lewis got a specimen of their language*[4]

Fire
water
earth
air
wind
sky
sun
moon
star
light
darkness
day
night
heat
cold
smoak
cloud
fog
rain
snow
hail
ice
frost
dew
rain-bow
thunder
lightning
yesterday
to-day
to-morrow
day
a month
a year
spring
summer
autumn
winter
a man
a woman
belly
back
side
bubby
nipple
thigh
leg
foot

toe
skin
nails
bone
blood
life
death
food
neat
fat
learn
bread
Indian-corn
milk
egg
a house
the mammoth
buffalo
elk
deer
moose
bear
wolf
panther
wind-cat
pole-cat
fox
monax
beaver
raccoon
opossum
to-day
to-morrow
a day
a month
a year
spring
summer
autumn
winter
a man
a woman
a boy
a girl
a child
father
mother

brother
sister
husband
wife
son
daughter
the body
the head
the hair
the beard
the face
an eye
the nose
the cheek
chin
lip
mouth
tooth
tongue
ear
neck
arm
wrist
hand
finger
moose
bear
wolf
panther
wind-cat
pole-cat
fox
monax
beaver
raccoon
opossum
hare
squirrel
flying-squirrel
ground-squirrel
mole
a bird
an eagle
hawk
owl
turkey
swan

wild-goose
duck
turkey-buzzard
raven
crow
black-bird
crane
pigeon
dove
pheasant
partridge
mocking-bird
red-bird
snake
lizard
butterfly
fly
fish
frog
mulberry
a vine
tobacco
joy
sorrow
one
two
three
four
five
six
seven
eight
nine
ten
eleven
twelve
thirteen
fourteen
fifteen
sixteen
seventeen
eighteen
nineteen
twenty
twenty-one
thirty
forty

fifty
sixty
seventy
eighty
ninety
a hundred
two hundred
three hundred
four hundred
five hundred
six hundred
seven hundred
eight hundred
ugly
sick
brave
cowardly
wise
foolish
I
you
he
she
they
this
that
to eat
to drink
to sleep
to laugh
to cry
to sing
to whistle
to smell
to hear
to see
to speak
to walk
to run
to stand
to sit
to lie down
to smoke
a pipe
to love
to hate
to strike

to kill
to dance
to jump
to fall
to break
to bend
yes
no
gold
silver
copper
a stone
wood
gum
a mountain
hill
valley
sea
lake
pond
river
creek
a spring
grass
a tree
pine
cedar
sycamore
poplar
ash
elm
beech
birch
maple
oak
chestnut
hiccory
walnut
locust
mulberry
a vine
tobacco
joy
sorrow
one
two
three

four
nine hundred
a thousand
white
black
green
blue
yellow
red
good
bad
large
small
high
low
broad
narrow
old
young
new
hard
soft
sweet
sour
better
hot
cold
dry
wet
strong
weak
pretty
ugly
sick
brave
cowardly
wise
foolish
I
you
he

Note: there is no explanation why Jefferson's list repeats some words.

APPENDIX SEVEN

THE TRAGIC LOSS

AS NOTED EARLIER, MERIWETHER LEWIS collected vocabularies for Thomas Jefferson whenever possible. Even though it must have been awkward and required many hours of tedious translation, Lewis faithfully carried out these instructions. Years later President Jefferson praised his diligence.

> *Capt. Lewis was instructed to take those of every tribe. . . . He was very attentive to this instruction, never missing an opportunity of taking a vocabulary.*[1]

Having for many years collected samples of Native American languages, Jefferson described in 1816 what he hoped to do with them:

> *the intention was to publish the whole, and leave the world to search for affinities between these and the languages of Europe and Asia.*[2]

President Jefferson was preparing to publish his research but changed his mind when he received the new vocabularies that Lewis brought back from the expedition.

> *I have now been thirty years availing myself of every possible opportunity of procuring Indian vocabularies to the same set of words . . . I had collected about 50. and had digested most of them in collateral columns and meant to have printed them the last year of my stay in Washington. But not having yet digested Capt. Lewis's collection, nor having leisure then to do it, I put it off till I should return home.*[3]

Jefferson shipped his belongings back to Charlottesville by water after leaving the White House.

> *The whole, as well digest as originals were packed in a trunk of stationary & sent round by water with about 30. other packages of my effects from Washington.*[4]

Unfortunately, this was a terrible mistake.

> *and while ascending James river, this package, on account of it's weight & presumed precious contents, was singled out & stolen.*[5]

After opening the trunk and finding nothing of value, the thief got rid of the evidence.

> *The thief being disappointed on opening it, threw into the river all it's contents of which he thought he could make no use. Among these were the whole of the vocabularies. Some leaves floated ashore & were found in the mud; but these were very few, & so defaced by the mud & water that no general use can ever be made of them.*[6]

Jefferson suppressed the emotion he must have felt after learning that this priceless collection of languages had been thrown away. He did add one sentence that indicates he knew that this body of work, which had taken him three decades to accumulate, was irreplaceable.

> *Perhaps I may make another attempt to collect, altho' I am too old to expect to make much progress in it.*[7]

APPENDIX EIGHT

ELK OR NO ELK

LEWIS AND CLARK ENCOUNTERED AN ODD distribution of elk along the lower Columbia. In the eighteen days they spent along the northern shore of the river they did not see a single elk. However, once they crossed over to the south side they saw hundreds and killed more than thirty-two of them in the first eighteen days.[1] Why were all the elk on the south side of the river?

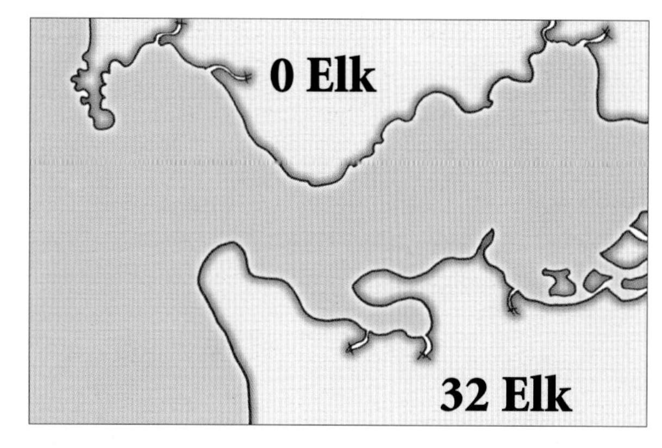

In the eighteen days Lewis and Clark spent along the northern shore they never saw and an elk; however, in the first eighteen days along the southern shore they killed more than thirty-two and saw hundreds more.

First, it is important to point out that the climate, vegetation, and landscape are practically identical in the two areas. Both sides of the river have thousands of acres of wetlands, bays, hills, and valleys. What's more, both areas support healthy elk populations today.

Where had the elk on the north side of the Columbia gone in 1805? Had they migrated farther north? Were they extremely well hidden? Or were the hunters simply unlucky?

Any of these explanations could be correct; however, there is another possibility, which requires looking northward several hundred miles to the tribes inhabiting that coastal region.

Along the Northwest Coast, above Vancouver Island, hundreds of thousands of native people, among them the Tlingit, Haida, Bella Coola, Kwakiutl, and Nootka, lived in large, spacious villages and enjoyed considerable wealth, a high level of artistic expression, and a complex social structure. Unfortunately, they often had to defend themselves from others who attempted to

A highly decorated Tlingit house with totem (c.1899) shows the unique abstract art style developed by Northwest indigenous people.

Interior of Haidia house c. 1890 shows extravagant splendor, craftsmanship, and high art style that existed along the Northwest Coast.

steal their riches or to crowd into their territory. Warfare among the tribes was common.[2]

Much like the knights of Europe's Middle Ages, the fighting men of these northern tribes dressed in heavy protective armor before they went into battle. However, their armor was made not of metal but of pieces of leather quilted together in such a manner that arrows could not penetrate it.[3] In fact, this armor was so efficient that Russian traders complained it would even stop their musket balls.[4]

The highest quality leather for making this armor apparently came from elk, and the greatest population of these animals was found far to the south, around Puget Sound and the Columbia River. Obviously, the northern tribes lived out of the range of these animals; however, the extensive trade networks that moved goods up and down the coast from village to village supplied whatever they needed. The northward flow of elk hides from the Columbia River had been going on for hundreds, if not thousands, of years.

Fur traders quickly learned about this demand for elk hides and immediately recognized that using their ships to transport bulky, dried hides from the Columbia to the northern tribes was good business. The elk hides were relatively inexpensive and could be readily traded for valuable sea otter pelts. In fact, elk hides became the principal reason fur traders came to the Columbia.

William Sturgis, who was unquestionably the leading authority and participant in the sea otter trade along the Northwest Coast, said the Columbia River was visited "chiefly to obtain some articles, which are again sold to the Indians on the more northerly part of the Coast."[5] In 1804, one year before Lewis and Clark's arrival, Capt. Sturgis purchased 145 elk hides from the Chinooks.

This enormous shipment reflects only a small portion of the elk hides being removed from the Columbia and Willamette valleys. The records from sailing ships describe a steady stream of hides leaving the river. In 1793, Captain Roberts brought his ship, the *Thomas Jefferson,* into the Columbia and purchased 124 elk hides for trade to northern people. In 1795-96, Captain Bishop of the *Ruby* wintered in the Columbia and purchased two hundred elk hides.[6]

Lewis and Clark were well aware that elk was a major trade item. On November 18, when passing the anchorage at Cape Disappointment, Clark described the trade between the Indians and the white men:

> *this rock Island is Small and at the South of a deep bend in which the nativs inform us the Ships anchor, and from whence they receive their goods in return for their peltries and Elk Skins* [7]

Lewis is more specific about what is exchanged:

The persons who usially visit the enterence of this river for the purpose of traffic or hunting, I believe is either English or Americans . . . This traffic on the part of the whites Consist in vending, guns . . . brass tea kettles, Blankets . . . Knives Beeds & Tobacco . . . for those they receive in return from the nativs Dressed and undressed Elk Skins, Skins of the Sea otter, Common Otter, beaver[8]

Both men specifically mention elk hides, which leaves little doubt what the fur traders were buying. In fact, Lewis and Clark were approached on numerous occasions by Indians who tried to sell them hides.

2 others Canoes Came one from the War-ci-a-cum Village, with three Indians . . . Those people brought with them Some Wapto roots, mats made of flags, & rushes, dried fish and Some fiew She-ne-tock-we roots & Dressed Elk Skins, all of which they asked enormous prices for, particularly the Dressed Elk Skins[9]

Some fur traders didn't bother purchasing elk from the natives. Instead, they landed and did their own hunting. Lewis and Clark were told about one particular captain named Davidson:

Davidson . . . Visits this part of the Coast and river in a Brig for the purpose of Hunting the Elk returns when he pleases he does not trade any, Kills a great maney Elk[10]

Clark observed that the Chinooks owned good firearms; perhaps these had been given to them by fur traders for the express purpose of obtaining elk hides. If this were the case, the Chinook hunters could have significantly reduced the elk populations around their villages in just a couple of years, especially with men like Davidson also hunting elk.

Elk hides from Puget Sound and the Columbia River were highly valued further north among the Tlingits, who called them "clemmels," the Haida, who referred to them as "clemens or clamons," and the Euro-Americans, who called them "clammels." Many elk hides were shipped north.

All the ships anchored along the northern shore of the Columbia, so logically that is where the first elk were hunted. The southern shore where Lewis and Clark found so many elk was relatively inaccessible to sailors; ships could not safely cross, and there was no protected anchorage.

The fur traders' ships had a huge capacity for cargo, and this, combined with their insatiable yearning for money, may well have resulted in a decimation of the elk populations. We will never know exactly why Lewis and Clark saw no elk along the north shore of the Columbia, but the northern tribes' demand for elk hides for their armor might be the explanation.

APPENDIX NINE

YORK AND SACAGAWEA

LEWIS AND CLARK DID NOT WASTE INK. THIS is especially true when it comes to writing about members of the party.

The men are rarely mentioned unless they are sick, hurt, have hunted successfully, or have been sent out on a special mission. In other words, the names of the men who paddled the canoes, set up camp, butchered the game, stood watch, chopped firewood, carried water, loaded the canoes, and stayed healthy rarely appear in the journals.

This is true of York and Sacagawea as well; their names appear only sporadically in the journals. However, during the month in the lower Columbia mentions their names frequently. In fact, they become major players in this history. York is described on five different occasions; Sacagawea is mentioned on seven separate days.

The first reference to York is when he returns successfully from hunting on November 16:

> *our hunters and fowlers killed 2 Deer 1 Crain & 2 Ducks, and my man York killed 2 geese and 8 Brant* [1]

On November 18, Clark brings York along with him on his excursion:

> *A little cloudy this morning I Set out with 10 men and my man York to the Ocian by land* [2]

On November 24, York participates in deciding where to camp for winter:

> *York (first choice) Examn other side (second choice) lookout* [3]

Clark sends York out fishing on December 2:

> *I despatched 3 men to hunt and 2 and my Servent in a Canoe to a Creek above to try & Catch Some fish* [4]

On December 4, York is sent out to retrieve elk meat with Sergeant Pryor, which we learn when he gets lost on their return.

> *Serjt Pryors joined us except my man york, who had Stoped to rite his load and missed his way . . . I delayed about half an hour before York Came up, and then proceeded around this Bay* [5]

Sacagawea is first mentioned on November 8 when she apparently becomes seasick.

> *The Swells were So high and the Canoes roled in Such a manner as to cause Several to be verry Sick. Reuben fields, Wiser McNeal & the Squar wer of the number* [6]

Clark makes a brief comment about her being angry at him on November 13, but we do not know why. It appears that he is talking about wappatoe roots, and since they were extremely hungry and without food at this moment perhaps she is upset that he didn't purchase several bushels of them.

> *Squar displeased with me for not [Sin?] Wap-to a excellent root which is rosted and tastes like a potato* [7]

Sacagawea becomes a central figure in the purchase of a sea otter pelt on November 20, when she gives up her belt of blue beads:

at length we precured it for a belt of blue beeds which the Squar – wife of our interpreter Shabono wore around her waste[8]

The following day she is repaid.

we gave the Squar a Coate of Blue Cloth for the belt of Blue Beeds we gave for the Sea otter Skins [9]

On November 24, she expresses her opinion about where she thinks the party should build a winter camp.

Janey in favour of a place where there is plenty of Potas [10]

On November 30, she gives Clark a piece of bread she had been saving for her son.

The Squar gave me a piece of Bread to day made of Some flower She had Cearfully kept for her child [11]

On December 3, Clark tells us that she knew a way to extract an unexpectedly large amount of grease from bones.

after eateing the marrow out of two Shank bones of an Elk, the Squar choped the bones fine boiled them and extracted a pint of Grease, which is Superior to the tallow of the animal [12]

York and Sacagawea soon fade into the background. During their months at Fort Clatsop, they are mentioned on only a couple of occasions, no more frequently than any of the other members of the party.

APPENDIX TEN

THE CHINOOK

This is not an official history of the Chinook people. It is, however, an attempt to briefly explain who the Chinooks were and where they are today, from my standpoint as a life-long neighbor in their ancestral land. The Chinook nation has yet to publish their history, which has been passed down orally from generation to generation; when they do, we will have the first accurate account of their story.

The Chinook people occupied and controlled a region from the ocean up the Columbia River for a distance of two hundred miles. Exactly how long the Chinook resided here is unknown. Tribal legends place them at the mouth of the Columbia River at the time of creation: archeologists find evidence of village sites that date back several thousand years. The precise number of centuries is impossible to ascertain; however, the Chinook had been here long enough to spread their language, culture, and influence throughout a vast area.

Their first contact with Europeans came sometime in the eighteenth century. There is evidence that suggests shipwrecked sailors were washed ashore. However, this early contact was sporadic and had very little impact on the Chinook way of life, especially when compared to what came next. An epidemic of smallpox swept through the coastal region some time around 1776. How many people died from the invisible killer is unknown but estimates of from 30 to 50 percent of the entire population are considered conservative. Chinook oral history suggested a death rate of 90 percent, which agrees with some modern demographic estimates. The precise number of deaths does not really matter; what is important to understand is that the Chinook nation was severely decimated.

Sixteen years later, Captain Robert Gray sailed his ship *Columbia Rediviva* into the Columbia River and mapped its location. After him, hundreds of others came to the river. Another smallpox epidemic swept through the region in 1800. Lewis and Clark came overland in 1805. More fur traders and sailors followed them.

The Astor party arrived in 1811 and built a trading post along the southern shore of the river, which remained an important trading center for the next dozen years. A great deal of interaction occurred between the Chinook people and the British/Canadian/French settlers. Some men married Chinook women and fathered children. The isolation combined with trade brought these people together.

By the 1830s and 1840s settlers were streaming into the Oregon Territory. Gold in California brought thousands more into the area, and with the foreigners came wave after wave of devastating disease. Malaria, typhoid, whooping cough, measles, and diphtheria swept through the region and took a deadly toll on all Northwest tribes. Trade routes were interrupted and economies shattered. Chinook power, wealth, and prestige diminished.

As more and more Euro-Americans arrived, they crowded in close to the Chinooks in order to gain access to salmon and shellfish. This brought more disease to the remaining populations. Unoccupied villages were leveled and cemeteries vandalized, with many Chinook skeletons sent as curiosities back east. Reduced by disease and poverty, the tribe that had once controlled the entire lower Columbia and a large share of

Northwest commerce were now no longer capable of defending themselves against a handful of squatters.

Aware that the Americans would continue to arrive in ever-increasing numbers, the Chinook signed a treaty in 1851 with government Indian agent Anson Dart. They generously agreed to give up their lands to the American government as long as they could keep their traditional village sites, fishing grounds, and cemeteries. Unfortunately, this government agent was unaware that in Washington, D.C., a new plan had been devised by Congress. The first treaty was never ratified; another treaty in 1855 was presented to the Chinook telling them to leave the Columbia River and move a hundred miles north to the lands of their historic competitors and adversaries, the Quinaults. The Chinook refused to sign this treaty; they had no intention of leaving the land of their ancestors along the Columbia River.

Meanwhile, more homesteaders poured into the area. Timber and salmon became boom industries. Towns rose up where villages had once stood. Railroads, sawmills, and canneries sprouted along the shore, cutting the Chinook off from water. Racial hatred was leveled towards the Indians who stood in the way of these resources, and this forced many Chinook to leave for the safety of other reservations. The Fourteenth and Fifteenth Amendments, abolishing slavery and granting civil rights, did not extend these rights to the indigenous peoples of North America.

The Chinooks were reduced to non-voting non-citizens in the land where they had lived for thousands of years. Some part-blood Chinook, who successfully concealed their heritage, blended in with the newly arrived Americans. Rumors of the tribe's extinction were widely publicized.

Citizenship for Indians came in 1924 with the Snyder Act, which gave them legal access to the government. Twelve years later in 1936, the Indians of Washington State received the right to vote. After World War II, in which many Native Americans distinguished themselves in combat, America had new enemies to focus its attention on, and the discrimination against Indians fell to an all-time low.

Desiring to obtain official recognition from the United States Government, the Chinook filled out the paperwork, documented their genealogies to prove that they were the direct descendants of the people who had conversed with Lewis and Clark, and submitted their application to the Bureau of Indian Affairs. What began as a process that seemed perfectly logical soon became embroiled in red tape. The Bureau of Indian Affairs rejected their application. The Chinook made the requested changes and resubmitted. Again it was rejected. Thus began the process of application and rejection that continues to this day.

There is a salmon named chinook; there is a chinook wind, a river, and a town; a Chinook helicopter, motor-home, and casino. However, in the eyes of the United States government, the people who call themselves Chinook do not exist.

ENDNOTES

PROLOGUE (XVI)

1. Jackson, Donald, ed., *Letters of the Lewis and Clark Expedition with Related Documents 1783-1854*. Vol.2, 2d ed., Urbana: University of Illinois Press, 1978, 654-655.

CHAPTER 1: THE FINAL MILES (1-19)

November 7

1. Moulton, Gary E., ed., *The Journals of the Lewis & Clark Expedition*. Lincoln: University of Nebraska, 1983-2001. 6:11. On November 3, Indians told Lewis and Clark that in two days of travel they would see three ships. Supposedly, these ships would have been in the river very near the ocean. Clark writes: "we met 2 Canoes, of Indians 15 in number who informed us they had Seen 3 Vestles 2 days below us."
2. Moulton, *Journals*, 6:50. It is not clear whether Lewis and Clark carried a copy of George Vancouver's map of the mouth of the Columbia River; however, Clark makes a reference to it on November 15, which suggests that they had closely studied it. "I could not See any Island in the mouth of this river as laid down by Vancouver."
3. This ridge and the final curve in the river are both visible from Vista Park in the town of Skamokawa, Washington.
4. Today this estuary is called the Lewis and Clark National Wildlife Refuge. Each winter well over 55,000 swan, geese, and ducks converge on this 35,000 acre site.
5. Scottish born Alexander MacKenzie had crossed the Northern Territory on two occasions prior to Lewis and Clark. By using bark canoes and local natives as crew members and guides, he first reached the Arctic Ocean in 1789. On his second attempt, MacKenzie steered southward from Lake Athabasca, descended the Peace River, and reached the Pacific Ocean below present-day Ketchikan, Alaska. The year was 1793.

Canoes

1. Moulton, *Journals*, 5:234-236, 5:243-248.
2. Ibid., 5:327-328.

Mouth of the Columbia River

1. Gibbs, James A., *Pacific Graveyard: A Narrative of Shipwrecks Where the Columbia River Meets the Pacific Ocean*. Portland: Binford & Mort Publishing, 1993.
2. Ross, Alexander, *Adventures of the First Settlers*. London: Smith Elder and Co., 1849, and Readex Microprint Corporation, 1966, 56-57.

November 8

6. Moulton, *Journals*, 5:333. Many of Lewis and Clark's party could not swim. For example, when passing down the treacherous waters near the "Narrows," the men who could not swim were assigned the job of portaging baggage along the shore while the good swimmers brought the canoes through the rapids.

   Lewis and Clark kept the canoes near the shoreline, and it is possible they did so for safety. If a canoe were to overturn it would be easier to rescue men and supplies.
7. The current of the Columbia River varies its speed during an ebb tide; however, it is not unusual for the river to flow at five to seven miles per hour. If Lewis and Clark paddled at five miles per hour and the river flowed at the same speed, they would have advanced downriver at a very quick rate. The source of this information concerning the Columbia's speed during ebb tide is from the Columbia River Bar Pilots.
8. Moulton, *Journals*, 11:389-390. There was a sense of uncertainty in the minds of Lewis and Clark, based on the conflicting information the Indians had given them. Joseph Whitehouse writes: "The Indians both at this, & the other Indian village that we passed this day, made signs to us that there were vessells lying at the Mouth of this River. Some of them signed to us that the Vessells were gone away from it."

Changing Clothes

1. Merk, Frederick, ed., *Fur Trade and Empire: George Simpson's Journal*. Cambridge: Harvard University Press, 1931, 64.
2. Ibid., 64.
3. Durry, ed., *Where Wagons Could Go; Narcissa Whitman & Eliza Spalding*. Lincoln: University of Nebraska Press, 1997. 92-93.
4. Moulton, *Journals*, 6:47, 5:342-343.

Rumors of White Men

1. Moulton, *Journals*, 5:89.
2. Ibid., 5:226.
3. Ibid., 5:235.
4. Ibid., 5:255.
5. Ibid., 10:159.
6. Ibid., 11:370.
7. Ibid., 5:351.
8. Ibid., 6:11.
9. Ibid., 6:15.
10. Ibid., 6:27.

November 9

9. Did Lewis and Clark understand tidewater? This is a difficult question to answer. Neither man grew up near the ocean, so it is impossible to know where they would have gained the experience; and since most of their men came from farming and frontier families, it is unlikely they would have ever seen it either. We will probably never know the answer, but what we do know is that in the first eighteen months of their journey they never had to think about tidewater. They could beach their canoes in the evening, eat dinner, and go to bed. In the morning the water level would be unchanged.

   Their first introduction to tidewater occurred on November 2 when the party passed below the Great Cascades. Clark noticed the level of the river moved up and down nine inches; further downriver on November 4, when they were still one hundred miles from the ocean, the tide rose and dropped only three feet. Since they were in their canoes all day, except for stopping to eat, they would have barely felt the effect of the rising river.

   However, their present situation brought them under the full influence of the ocean. It was an entirely new predicament and from all appearances, they did not fully anticipate its effects.
10. On November 9, 1805, the full moon rose at 6:46 P.M. Even through thick clouds, the full moon sheds enough light along the Columbia River to discern between shoreline and water. In fact, this moonlight reflects off the breaking waves and makes them clearly visible. However, the shoreline itself, with its rocks and driftwood, remains as black as the inside of a chimney, and it is practically impossible to take a step without tripping.
11. According to the Army Corps of Engineers, it takes 800 million cubic yards of water to raise the Columbia River from low tide to high tide. Translated into a comprehensible measure, this means 160,800,000,000 gallons of ocean water rush into the Columbia in the six or seven hours between low tide and high tide. This is equivalent to 24,700,000,000 gallons of ocean water every hour, or 411,000,000 gallons every minute, or 7,000,000 gallons every second.
12. Moulton, *Journals*, 6:38. Clark. writes: "Some of the party not accustomed to Salt water has made too free a use of it on them it acts as a pergitive."

When Indians Camp

1. Moulton, *Journals*, 6:262-263.

Windchill

1. Glover, Thomas J., *Pocket Ref: Air and Gases*. Littleton: Sequoia Publishing, Inc. 1995.

CHAPTER 2: STRUGGLE AROUND POINT DISTRESS (20-49)

November 10

1. Moulton, *Atlas of the Lewis and Clark Expedition*, 1:89. On some maps Clark shows their approximate canoe route by drawing a faint dotted line. Whenever we see this in the lower river, it inevitably follows close along the shoreline.
2. Moulton, *Journals*, 6:104. "The emence Seas and waves which breake on the rocks & Coasts . . . roars like an emence fall at a distance."
3. The collision between the ebbing current and the flooding current goes largely unnoticed; however, when we realize that billions of tons of water hit head-on, it is easy to understand how this would logically bring the river to a momentary standstill. *High water slack* occurs at the end of the floodtide, moments before the river current begins to ebb; *low water slack* occurs at the end of the ebb tide. Sometimes, due to excessive wind or a freshet, the current does not stop; at other times, the still water exists for dozens of minutes. When *slackwater* does occur, it is the perfect time for mariners and fishermen to move their boats or position equipment.

Winter Storms

1. Oregon Weather Summary, Oregon Climate Service, Oregon State University.

   Questions about the weather Lewis and Clark encountered are frequently asked, especially regarding the rainfall. Their descriptions of the winter of 1805-1806 sound unbelievable. However, the rain Lewis and Clark experienced still falls on this region. Rainfall for the months of November and December 2001, shown in the chart below, was considered typical, not unusual.

November 2001 — Rainfall in inches

Day	in.	Day	in.	Day	in.	Day	in.
12	2.20	17	0.04	22	3.01	27	0.06
13	2.60	18	0.00	23	1.10	28	1.48
14	5.94	19	0.87	24	0.16	29	2.52
15	3.23	20	1.22	25	0.13	30	0.94
16	0.57	21	1.34	26	0.21		

December 2001 — Rainfall in inches

Day	in.	Day	in.	Day	in.	Day	in.
01	3.52	07	0.61	13	1.71	19	1.34
02	0.71	08	0.15	14	2.75	20	0.21
03	0.41	09	0.77	15	0.43	21	0.11
04	0.88	10	1.16	16	4.72	22	0.05
05	0.95	11	0.50	17	2.05	23	0.00
06	0.45	12	0.60	18	0.44		

Source: Naselle Salmon Hatchery, Naselle, Washington.

Indian Tribes along the Lower Columbia

1. Moulton, *Journals*, 6:202.
2. Ibid., 6:75.
3. Ibid., 6:121.
4. Ibid., 6:119.
5. Ibid., 6:168.

November 11

4. Moulton, *Journals,* 6:410-414. Clark tells us the Indians sold them "Red Charr." Burroughs, in his well-known book, *The Natural History of the Lewis and Clark Expedition*, mistakenly

identifies them as "Sock-eye Salmon." Unfortunately, it would have been practically impossible for Lewis and Clark to have seen a sock-eye salmon in November.

The sock-eye salmon enter the Columbia River in the months of May and June. These fish are bright silver when they swim in from the ocean. They migrate up river in July and August and enter large lakes hundred of miles from the ocean. It is there that they turn red and begin their reproductive cycle in September. By October the sock-eye are near the end of their lives; all adult sock-eye salmon are dead before November.

The identity of these "Red Charr" is unknown. Some have suggested they were the common coho salmon, which can take on a rusty-red color. However, it seems Clark would have used the term *red salmon* if they had been salmon. He specifically uses the term "Charr" which suggests he was seeing some characteristics in the fish that resembled the Atlantic char. The distinctive difference is in their spots; salmon have *dark spots* whereas char have *white spots*.

Possibly the Indians had managed to catch *bull trout* which also live in the Columbia River, turn bright red, and being of the char family, have the characteristic white spots. Bull trout have been photographed in schools, crowded into streams, which might explain how the Indians were able to catch thirteen fish at once.

It is also possible that the Indians had caught some other species of Char that is now extinct. Long before biologists had an opportunity to study fisheries in the Columbia, many runs of fish had been decimated. It is possible that inhabitants of the region who arrived here after Lewis and Clark strung nets from shore to shore across small streams and caught every last fish that entered that particular river. That practice was not uncommon even in the 1920s and 1930s. If this had been done back in the 1840s and 1850s when the Columbia salmon were starting to be exploited for commercial use, it is possible, even likely, that numerous species of fish were pursued to extinction.

Lewis and Clark describe both salmon and char in their journals and specifically separate those species. They tell us they see "Red Charr" in small streams, which is another indication the fish they called "Charr" was not sock-eye.

Elegant Canoes

1. Moulton, *Journals*, 6:262.
2. Ibid., 6:265.
3. Ibid., 6:262.
4. Ibid., 6:263.
5. Ibid., 6:263.
6. Ibid., 6:271-272.

November 12

5. Moulton, *Journals*, 6:198-199. We don't know many details about the Indian canoe Lewis and Clark used; however, they give us a few clues about it when it happens to get lost during the winter of 1805. Lewis writes, "this will be a very considerable loss to us if we do not recover her; she is so light that four men can carry her on their shoulders a mile or more without resting." This suggests the canoe weighed somewhere between two and three hundred pounds and probably measured somewhere around thirty-three feet in length.
6. Moulton, *Journals*, 5:323-324; 11:370. When Clark mentions "Pounded Salmon" he is referring to the dried salmon meat prepared by the Indians two hundred miles up river. He describes how the Indians processed this fish so that it was perfectly dehydrated and then stored it in long, narrow baskets, each weighing ninety pounds; it remained edible for two years. When soaked in water and heated, the reconstituted fish became a nutritious meal. Joseph Whitehouse says they purchased 1,440 pounds of this dried fish.

   This particular purchase of dried salmon was extremely fortunate. There are many days in the lower Columbia when the men would have suffered hunger if it had not been for this "pounded salmon."
7. The rainfall along the Pacific Northwest coast is unpredictable and varies widely from region to region. For example, the average rainfall in Cannon Beach, Oregon, between 1995 and 1999 was 110 inches per year. In those same years, forty miles to the north at Long Beach, Washington, the average was ninety-five inches of rain. It would seem logical that the heaviest amounts of rain fall near the ocean; however, this is not the case. The greatest amount of rainfall occurs in the valleys ten to twenty miles inland from the ocean. For example, one of the dampest places in all of Washington State is a valley that lies only fifteen miles east of Long Beach and several miles north of the Columbia River. The following rainfall was officially recorded in this depressingly wet community:

1995	1996	1997	1998	1999
132.7	132.2	136.5	137.5	144.3

   inches of rain per year

   Source: Washington State Salmon Hatchery, Naselle, Washington.
8. Moulton, *Journals*, 6:48, 6:52, 11:393. William Clark and Joseph Whitehouse both become confused about the exact day. It does not appear that they were copying each other because they mistake different days. On November 15, Clark mistakenly writes the date as the 16th: "The rainey weather Continued without a longer intermition than 2 hours at a time from the 5th in the morng. Untill the 16th is eleven days rain." Whitehouse dates his journal entry for November 13 correctly, but misdates the following day. He writes, "Wednesday Novemr. 13th" and on the following day he writes, "Thursday Novembr. 13th"

November 14

9. Moulton, *Journals*, 5:250-252, 343, 367, 371. The repairs to the canoes begin a day after they first launch them back in October. The men install braces and make numerous repairs in the following weeks. By the time they had reach the lower Columbia, these men know how to repair cracks, splits, and leaks of any description.
10. Moulton, *Journals*, 10:170. Gass tells us that Colter returned because his rifle broke. "About noon one of the 3 men who had gone in the canoe, returned having broke the lock of his gun."

"Those Scoundrals"

1. Moulton, *Journals*, 5:288, 317, 341, 342, 347.
2. Ibid., 5:305.
3. Ibid., 6:17-18.
4. Ibid., 6:18.
5. Ibid., 6:31.
6. Ibid., 5:323.

Chapter 3: The Arrival (50-69)

November 15

1. President Jefferson instructed Lewis to explore the Missouri River and whatever stream connected it to the Pacific Ocean. He did not say anything about exploring out into the ocean or along its coastline. That work had been started decades earlier by Captain Cook and was being carried out by Captain Vancouver and dozens of other fur trader/explorers who sailed along the shore.

   Upon passing around that final point of land and arriving in full view of the ocean, Clark could now describe the distance, direction, width, and depth of the rivers that connected the Mississippi with the Pacific.

Where are the Chinooks?

1. Swan, James G., *The Northwest Coast or Three Years' Residence in Washington Territory*. Seattle: University of Washington Press, 1997.
2. Franchere, Gabriel; *A Voyage To The Northwest Coast Of America*, ed. Milo Milton Quaife. Chicago: The Lakeside Classics, 92.
3. Ibid., p. 118.
4. Coues, Elliot, *The Manuscript Journals of Alexander Henry and of David Thompson, 1799-1814*. Minneapolis: Ross & Haines, Inc., 789.
5. Ibid., 837.

November 17

2. To find the precise location of significant points on the opposite side of the river, Clark first had to triangulate across using his compass. The data from this survey is found Moulton, *Journals*, 6:113-114. This data was turned into a map which is reproduced: Moulton, *Atlas of the Lewis and Clark Expedition*, 1:90. The inaccurcies in the shape of the southern shoreline reveal that this map was drawn before they had crossed the Columbia. Point Adams is shown incorrectly as an island; Lewis first discovers that it is attached to the mainland on December 1.
3. Moulton, *Journals*, 6:196. Clark repeatedly describes their short supply of trade goods. He writes: "Lewis examined our Small Stock of merchendize found Some of it wet and Dried it by the fire. Our merchindize is reduced to a mear handfull, and our Comfort, dureing our return next year, much depends on it."
4. Gibson, James R., *Otter Skins, Boston Ships, and China Goods: The Maritime Fur Trade of Northwest Coast, 1785-1841*. Seattle, University of Washington Press, 1992, 115-118; Suttles, Wayne, vol. ed., *Handbook of North American Indians: Vol.7, Northwest Coast*. Washington, DC: Smithsonian Institute, 1990, 119-124. Early fur traders to the Northwest coast wanted to immediately begin trading when they met natives with furs; however, they soon learned that was not proper. Before trading began there was a period of gift exchange. The natives would offer gifts and it was expected that they would receive something in return. Sometimes this exchange of gifts required two or three days before the natives were satisfied and felt it was time to begin business. This went against the nature of the fur traders, who wanted to purchase as many furs as possible in the shortest period of time, but they quickly learned that if they expected to do any business they had to follow the native custom.
5. Thwaites, Reuben Gold, ed., *Original Journals of the Lewis and Clark Expedition, 1804-1806*, Vol, 7, Jefferson's Letter to Lewis, June 20, 1803. Arno Press, 1969, 248. President Jefferson requested that Lewis estimate the population of the native people he encountered. Jefferson wrote: "renders a knolege of these people important, you will therefore endeavor to make yourself acquainted, as far as a diligent pursuit of your joureny shall admit, with the names of the nations & their numbers."
6. Ibid., 283-287. President Jefferson and his colleagues at the American Philosophical Society requested a wide range of ethnological information about the native people who resided in the western half of the continent. This information included details about such things as life expectancy, medicine, clothing, and observations on the division of labor and diet. A typical question was: "What is their most general diet, manner of cooking, time and manner of eating; and how doe they preserve their provisions?"

Trade Goods

1. Moulton, *Journals*, 6:205.
2. Gibson, *Otter Skins, Boston Ships, and China Goods*, "The Maritime Fur Trade of the Northwest Coast, 1785-1841," 215-217.
3. Ibid., 214-215.

Men Who Accompanied Clark
1. Moulton, *Journals*, 6:60, 62, 65.
Arrival Date
1. Moulton, *Journals*, 6:429.
2. Jackson, *Letters*, 1:329-330.
Indian Population
1. Jackson, *Letters*, 1:62.
2. Moulton, *Journals*, 6:61.
3. Ross, Alexander, *Adventures of the First Settlers*, 256.
Speculations on a Map
1. Thwaites, *Original Journal*, 3:234-235

Chapter 4: Clark's Excursion (70-93)
November 18
1. Coues, Elliott, *Henry & Thompson Journals*, Vol. II. Minneapolis: Ross & Haines, Inc., 763. Clark and his men must have noticed these names carved into the trunks of trees. In 1813 Alexander Henry came ashore at the cape and saw them.
Henry writes: "At 10 a. m. we went ashore to see the cape. Here were a party waiting at the old spot. The many letters engraved on trees near the spring gave us reason to suppose that this harbor had been much frequented by American vessels. Some names still legible were: H. Thompson, ship Guatimozin of Boston, Feb. 20th, 1804; ship Caroline of Boston, May 21st, 1804. There were several other inscriptions which, from the bark having overgrown the letters, and fire having passed, we were unable to make out."
2. The two waterfalls on the Columbia River were located seventy miles apart.The upper one Clark called "The Great Falls of the Columbia River" and became known as Celilo Falls. The lower fall was located at the Great Cascades which Clark called "Great Shute" or the "Grand Rapid."
The Anchorage
1. Moulton, *Journals*, 6:201.
2. Ibid., 6:205; Ruby, Robert H. and Brown, John A., *The Chinook Indians, Traders of the Lower Columbia River*. Norman: University of Oklahoma Press, 1976.115-116; Gibson, *Otter Skins*, 233-239. Slavery was common among the native people of North America. Even Sacagawea had been captured and sold. In the Columbia River Basin, slaves were captured or purchased from outside the region, then transported and sold to whomever could afford to own them. Some sailing ship logs indicate that fur traders became involved in this northwest slave trafficking. They purchased slaves from dealers, then transported them northward where they were later sold or traded for furs.
Moulton, *Journals*, 6:360. On February 28, 1806, a Clatsop man tried to sell a slave boy to Clark. Clark writes: "Kus ke lar a Clatsop man, his wife and a Small boy (a Slave, who he informed me was his Cook , and offered to Sell him to me for beeds & a gun) visited up today."
Ibid., 6:367. Clark adds: "The boy which this Indian offered to Sell to me is about 10 years of age. this boy had been taken prisoner by the Kil a mox from Some Nation on the Coast to the S. east of them at a great distance."
3. Moulton, *Journals*, 6:66.

November 19
3. In June 1805, when the party was still crossing Montana, several passages indicate that the men's clothes were wearing out. By June 23 it appears the men's shoes were worn out. Clark writes:"the men mended their mockersons with double Soles to Save their feet from the prickley pear."
4. The shape of the Cape Disappointment peninsula was radically altered when the North Jetty was constructed in 1914-1916. Sand accumulated behind the rock, which added thousands of acres of beach to the cape where before there had been ocean.
Tracking
1. Moulton, *Journals*, 5:37.
Walking
1. Moulton, *Journals*, 4:327-328.
Willapa Bay
1. Moulton, *Journals*, 6:70.
Where Are the Trees?
1. Moulton, *Journals*, 6:279.
2. Ibid., 6:42.
3. Ibid., 6:70.
4. Ibid., 6: 276-277.

November 20
5. Moulton, *Journals*, 6:201.
6. A close examination of Lewis and Clark's diet as revealed in the journals indicates that they were subsisting on geese, ducks, swans, and herons.
7. Moulton, *Journals*, 6:85.
Four Possible Winter Campsites
1. Moulton, *Journals*, 5:333.
2. Ibid., 5:335.
3. Ibid., 5:336.
4. Ibid., 6:11.
5. Ibid., 5:349.
6. Ibid., 5:349.
7. Ibid., 5:339.

Chapter 5: Unexpected Change of Plans (94-107)
November 21
1. Moulton, *Journals*, 9:242; 10:176; 11:397. Ordway, Gass, and Whitehouse all mention on November 21 the purchase of the sea otter pelts in exchange for Sacagawea's belt of blue beads. The actual purchase occurred on the day before, so one cannot help but wonder if it were the blue coat that brought this to their attention.
2. Merk, *Fur Trade and Empire: George Simpson's Journal 1824-1825*, 100-101; Franchere, *Voyage to the Northwest Coast of America*, 182-183; Coues, *The Manuscript Journals of Alexander Henry and of David Thompson, 1799-1814*, 836. When the Astor party and North West Company built permanent settlements at the mouth of the Columbia, the residents described the local prostitutes as young slave girls compelled by their owners to offer themselves to the foreigners, which because of disease often led to a miserable and early death. There exists some speculation about whether all prostitutes were slaves.
3. Moulton, *Journals*, 6:75.
4. Ibid., 6:73.
Blue Beads
1. Moulton, *Journals*, 5:371.
2. Ibid., 6:214-215.
3. Ibid., 6:205.
4. Ibid., 7:253.
Sick Men
1. Moulton, *Journals*, 6:239.
2. Ibid., 6:393.
3. Ibid., 6:418.

November 22
4. It is not unusual for extremely strong winds to hit the Northwest Coast in wintertime. Wind speeds of eighty to ninety miles per hour are common, and speeds of one hundred miles per hour have been recorded on many occasions. The Fort Canby State Park records state that in 1893 the wind exceeded 120 miles per hour and in 1923 the wind reportedly exceeded 160 miles per hour.

November 23
5. The environment near the seacoast is extremely harsh on iron. The salt water is carried by the wind and immediately attacks any iron or steel. Rust appears in a matter of days, whereas under normal conditions the appearance of rust would take weeks or months. Lewis and Clark's rifles, knives, tomahawks, and axes were undoubtedly beginning to turn rusty orange.
6. This is the only recorded occasion on which the entire party mark a campsite with their names. Clark says they write on alder trees, which were presumably somewhat mature at the time. Alder have a short life expectancy; sixty or eighty years is old. Even if no one had vandalized the site and the trees were left untouched, they would have all died and rotted before 1860.
"Great Higlers"
1. Ames & Maschner, *Peoples of the Northwest Coast*, 165-176.
2. Gibson, *Otter Skins, Boston Ships, and China Goods*, Chapter 6; Ross, *Adventures of the First Settlers*, 77-78.
3. Merk, *Fur Trade and Empire*; 96.
4. Moulton, *Journals*, 6:165.
Sailing Ships and Disease
1 Platt, Colin, *King Death, The Black Death and Its Aftermath in Late-Medieval England*. Toronto: University of Toronto Press, 1996, 19-20; Tuchman, Barbara, *A Distant Mirror; The Calamitous 14th Century*. New York: Ballantine Books, Chapter 5.
2. Ames & Maschner., *Peoples of the Northwest Coast*, 53-56.
3. Suttles, Wayne ed., *Handbook of North America Indians*, 135-148.
4. Ames & Maschner, *Peoples of the Northwest Coast*, 53-56.
5. Moulton, *Journals*, 6:285.
6. Ruby and Brown, *The Chinook Indians*, Chapter 11: "The Cold Sick."
7. Ames & Maschner., *Peoples of the Northwest Coast*, 53.
Jack Ramsay
1 Cox, Ross, *The Columbia River*. University of Oklahoma Press, 1957, 170-171.
2. Moulton, *The Journals*, 6:147-148.

Chapter 6: Another Change of Plans (108-119)
November 24
1 Moulton, *Journals*, 6:47; Ibid., 6:86. Clark observed that if they were to have freezing weather before they could replenish their supply of clothes, the men would be in serious trouble. "if we have Cold weather before we Can kill & Dress Skins for Clothing – the bulk of the party will Suffer verry much." Clark also refers to "our naked party dressed as they are altogether in leather."
2. Moulton, *Atlas of the Lewis and Clark Expedition*, 1:90. Clark had already drawn a map of the southern shore, based on survey measurements. If it had been his intention to cross the Columbia, it is hard to imagine he would have gone to the work of making the map.
3. Moulton, *Journals*, 6:44 & 6:171. Sacagawea confronted Clark and spoke her mind on two recorded occasions. The first was on, or around, November 13. Clark said she was "displeased" with him about something (possibly his failure to buy wappatoe roots). The second time she confronted him was on January 6. Clark was planning a trip to the ocean to get whale blubber and she was being left behind. "The last evening Shabono and his Indian woman was very impatient to be permitted to go with me, and was therefore indulged; She observed that She had traveled a long way with us to see the great waters, and that now that monstrous fish was also to be Seen, She thought it verry hard that She Could not be permitted."

These two examples do not prove anything conclusively, but they do suggest that Sacagawea did not always wait to be spoken to, and would tell Clark what she was thinking, even if it went against his plans.

Sextant Navigation

1. Jackson, *Letters*, 1:.61-62.
2. Jackson, *Letters*, 1:.49.
3. Moulton, *Journals*, 10:177.
4. Approximation from NOAA Coast Survey Charts.

Greatest Number of Deer

1. Moulton, *Journals*, 7:20.
2. Ibid., 7:22.
3. Ibid., 6:19.

My Man York

1. Moulton, *Journals*, 6:53.
2. Ibid., 6:105.

Sacagawea's Response

1. Moulton, *Journals*, 6:17.

"Consultation of the Men's Opinions"

1. Moulton, *Journals*, 2:501.
2. Ibid., 4:248.
3. Ibid., 8:302.

Chapter 7: In Search of Elk (120-139)

November 25

1. Moulton, *Journals*, 6:87. Clark says: "we Dined in the Shallow Bay on Dried pounded fish, after which we proceeded on."

November 26

Indian Burial

1. Moulton, *Journals*, 6:186.
2. Ibid., 6:186.

November 28

Hunters

1. Moulton, *Journals*, 6:200.

Tides

1. Moulton, *Journals*, 6:109.
2. National Oceanic & Atmospheric Administration, National Ocean Service, Silver Springs, Maryland.

Chapter 8: Lewis's Excursion (140-157)

November 29

1. The place the Indians pointed to was apparently near the tip of Point Adams, which was west, straight across the bay and four or five miles away. However, to get there, Lewis would have to follow close along the shore, which curved around and made the trip seven or eight miles long.

November 30

2. Clark describes the coots as "3 black Ducks with Sharp White beeks" but there is no mention of the men having eaten them.

"This Bread I eat with Great Satisfaction"

1. Moulton, *Journals*, 2:217.
2. Ibid., 5:189.

December 1

3. When I first read Lewis's brief description of these plants I was perplexed. Why would he be wasting this precious time writing short descriptions of local plants when he and his men should be out hunting? Finally it occurred to me that this was an indication that Lewis had decided to forget about building a winter camp near the ocean and instead built a camp farther upriver near the Cascades, as he had originally planned. If Lewis intended to remain in this vicinity all winter he would not spend this time describing the same plants he would be surrounded by for the next three months.
4. President Jefferson's request on June 20, 1803 specifically asked Lewis to notice "the soil & face of the country, it's growth & vegetable production" Lewis undoubtedly knew he had overlooked this plant community. He had collected two plants along the coast thus far. He might have been thinking about this obvious gap in his collection, and realized that Jefferson, Barton, and others would ask him about the plants growing near the ocean.
5. Lewis resumes his journal on January 1, 1806 after a silence of nearly three months, except for those three days at the end of November and first of December.

Botany

1. Moulton, *Journals*, 12:316; Dozleff, Eugene, *Shorelife of the North Pacific Coast*, 1983; Poiar & Mackinnon, *Plants of the Pacific Northwest*, 1994.

Chapter 9: Clark Awaits Lewis (158-175)

December 3

1. Moulton, *Journals*, 6:106.

"Hungry Men eat elk in Two Meals"

1. Moulton, *Journals*, 4:379.
2. Ibid., 6:172.
3. Ibid., 6:177.

December 5

2. Moulton, *Journals*, 10:180. Gass tells us: "Capt. Lewis and three of his party came back to camp; the other two were left to take care of some meat they had killed."
3. After spending several days looking for elk without any sign, Lewis would have been cautious about Drouillard's success. He would not have wanted to move the entire party near the ocean, have plenty of elk for several weeks, and then find themselves hungry because all the game had been killed. Lewis knew they needed a minimum of three elk every two days, and he would have had to make sure there were enough animals in the surrounding vicinity to make it worth their while before he brought the party there.

Chapter 10: A Winter Camp at Last (178-183)

December 6

1. Hicks, Steacy D., *Our Restless Tides: A Brief Explanation of the Basic Astronomical Factors which Produce Tides and Tidal Currents, 1-24*; Abell, Morrison, Wolf, *Realm of the Universe*, 44-48. Astronomical data.

December 7

2. Moulton, *Journals*, 6:278. On February 5, one of Lewis's men yelled and fired his gun to attract attention. "Late this evening one of the hunters fired his gun over the swamp of the Netul opposite to the fort and hooped. I sent sergt. Gass and a party of men over." They might have done the same to attract York's attention and guide him towards their canoe.
3. Ibid., 10:199. Gass reports the number of elk and deer shot at Fort Clatsop as: "I made a calculation of the number of elk and deer killed by the party from the 1st of Dec 1805 to the 20th of March 1806, which gave 131 elk, and 20 deer."

   Clark estimated the elk killed to be several dozen more than Gass's number.

Epilogue

1. Jackson, *Letters*, 1:319-320.

APPENDICES (188-209)

To See or Not To See

1. Gibbs, *Pacific Graveyard*, 39-48. Willingham, William F., *Army Engineers and the Development of Oregon*. U.S. Government Printing Office, 1992. 60-69.
2. Thwaites, *Original Journals*, 3:210.
3. Moulton, *Journals*, 6:103.

Fur Trade around the World

1. Gibson, *Otter Skins*, 84-90; Munford, James Kenneth, ed., *John Ledyard's Journal of Captain Cook's Last Voyage*. Corvallis: Oregon State University Press, 1963, 200.
2. Vaughn, Thomas and Holm, Bill. *Soft Gold: The Fur Trade & Cultural Exchange on the Northwest Coast of America*. Portland: Oregon Historical Society Press, 1990, 17-26.
3. Gibson, *Otter Skins*, 38-69.

Ship Repair

1. Howay, Frederic W., ed., *Voyages of the "Columbia" to the Northwest Coast, 1787-1790 and 1790-1793*. Portland: Oregon Historical Society Press, 1990, 170.
2. Moulton, *Journals*, 6:201.
3. Howay, *Voyages of the "Columbia,"* 356.

Jefferson's Letter to Lewis, 1803

1. Jackson, *Letters*, 1:61.

Jefferson's Letter of Credit

1. Jackson, *Letters*, 1:105-106.

Jefferson's Vocabulary List

1. Moulton; *Journals*, 5:287.
2. Ibid., 5:344-345.
3. Ibid., 5:345.
4. Ibid., 10:175.

Tragic Loss

1. Jackson, *Letters*, 2:611.
2. Ibid., 2:611.
3. Ibid., 2:465.
4. Ibid., 2:465.
5. Ibid., 2:465.
6. Ibid., 2:465.
7. Ibid., 2:465.

Elk or No Elk

1. Moulton, *Journals*, 6:30-123.
2. Smithsonian, 126-128.
3. Ibid., 218; Ames & Maschner, *Peoples of the Northwest Coast*, 195–218; Gibson, *Otter Skins*, 9-11, 205, 230-231.
4. Gibson, *Otter Skins*, 9-11.
5. Ibid., 230-231.
6. Ibid., 230-231.
7. Moulton, *Journals*, 6:66.
8. Ibid., 6:205.
9. Ibid., 6:146.
10. Ibid., 6:156.

York and Sacagawea

1. Moulton, *Journals*, 6:54.
2. Ibid., 6:65.
3. Ibid., 6:84.
4. Ibid., 6:105.
5. Ibid., 6:114.
6. Ibid., 6:35.
7. Ibid., 6:44.
8. Ibid., 6:73.
9. Ibid., 6:73.
10. Ibid., 6:84.
11. Ibid., 6:96.
12. Ibid., 6:107.

BIBLIOGRAPHY

Ames, Kenneth M. and Maschner, Herbert D.G., *Peoples of the Northwest Coast: Their Archaeology and Prehistory*. London: Thames and Hudson, Ltd., 1999.

Betts, Edwin Morris and Bear, James Adam, Jr., eds., *The Family Letters of Thomas Jefferson*. Charlottesville: University Press of Virginia, 1995.

Burroughs, Raymond Darwin, ed., *The Natural History of the Lewis and Clark Expedition*. East Lansing: Michigan State University Press, 1995.

Busch, Briton C. and Gough, Barry M., eds., *Fur Traders from New England: The Boston Men in the North Pacific, 1787-1800; The Narratives of William Dane Phelps, William Sturgis & James Gilchrist Swan*. Spokane: The Arthur H. Clarke Company, 1997.

Coues, Elliott, ed., *The Manuscript Journals of Alexander Henry Fur Trade of the Northwest Company and of David Thompson Official Geographer and Explorer of the same Company, 1799-1814*. Vol. 1 & 2: "The Saskatchewan and Columbia Rivers." Minneapolis: Ross & Haines, Inc.

Cox, Ross, *The Columbia River: Or scenes and adventures during a residence of six years on the western side of the Rocky Mountains among various tribes of Indians hitherto unknown; together with "A Journey across the American Continent."* Norman:University of Oklahoma Press, 1957.

Franchere, Gabriel, *A Voyage to the Northwest Coast of America*. Chicago; Lakeside Press, 1954, and New York: Redfield, 1854.

Gibbs, James A., *Pacific Graveyard: A narrative of shipwrecks where the Columbia River meets the Pacific Ocean*. Portland: Binford & Mort Publishing, 1993.

Gibson, James R., *Otter Skins, Boston Ships, and China Goods: The Maritime Fur Trade of Northwest Coast, 1785-1841*. Seattle, University of Washington Press, 1992.

Glover, Thomas J., *Pocket Ref: Air and Gases*. Littleton: Sequoia Publishing, Inc., 1995.

Howay, Frederic W., ed., *Voyages of the "Columbia" to the Northwest Coast, 1787-1790 and 1790-1793*. Portland: Oregon Historical Society Press, 1990.

Jackson, Donald, ed., *Letters of the Lewis and Clark Expedition with Related Documents 1783-1854*. Vol. 1 & 2, 2d ed., Urbana: University of Illinois Press, 1978.

Hicks, Steacy D., *Our Restless Tides: A Brief Explanation of the Basic Astronomical Factors Which Produce Tides and Tidal Currents, 1-24*.

Koch, Adrienne and Peden, William, *The Life and Selected Writings of Thomas Jefferson*. New York: Random House, Inc., 1972.

Lyon, E. Wilson, *Louisiana in French Diplomacy, 1759-1804*. Norman: University of Oklahoma Press, 1934.

Mallory, Mary, *"Boston Men" on the Northwest Coast: The American Maritime Fur Trade, 1788-1844*. Kingston: The Limestone Press, 1998.

Merk, Frederick, ed., *Fur Trade and Empire: George Simpson's Journal*. Cambridge: Harvard University Press, 1931.

Moulton, Gary E., ed., *The Journals of the Lewis and Clark Expedition*. 13 vols. Lincoln: University of Nebraska Press, 1983-2001.

Munford, James Kenneth, ed., *John Ledyard's Journal of Captain Cook's Last Voyage*. Corvallis: Oregon State University Press, 1963.

Peterson, Merrill D., ed. *Thomas Jefferson: Writings*. New York: Literary Classics of the United States, Inc., 1984.

Platt, Colin *King Death, The Black Death and Its Aftermath in Late-Medieval England*. Toronto: University of Toronto Press, 1996.

Ross, Alexander, *Adventure of the First Settlers on the Oregon or Columbia River: being a Narrative of the Expedition Fitted out by John Jacob Astor, to Establish the Pacific Fur Company; with an Account of Some Indian Tribes on the Coast of the Pacific*. London: Smith Elder and Co., 1849, and Readex Microprint Corporation, 1966.

Ruby, Robert H. and Brown, John A., *The Chinook Indians, Traders of the Lower Columbia River*. Norman: University of Oklahoma Press, 1976.

Suttles, Wayne, vol. ed., *Handbook of North American Indians: Volume 7, Northwest Coast*. Washington, DC: Smithsonian Institute, 1990.

Swan, James G., *The Northwest Coast or Three Years' Residence in Washington Territory*. Seattle: University of Washington Press, 1997.

Thwaites, Reuben Gold, ed., *Original Journals of the Lewis and Clark Expedition, 1804-1806*. Arno Press, 1969.

Tuchman, Barbara W., *A Distant Mirror: The Calamitous 14th Century*. New York: Ballantine Books, 1979.

Vaughn, Thomas and Holm, Bill, *Soft Gold: The Fur Trade & Cultural Exchange on the Northwest Coast of America*. Portland: Oregon Historical Society Press, 1990.

Whitman, Narcissa and Spalding, Eliza, *Where Wagons Could Go*. Lincoln: University of Nebraska Press, 1997.

Willingham, William F., *Army Engineers and the Development of Oregon*. U.S. Government Printing Office, 1992.

Wilson, Douglas L. and Stanton, Lucia, *Jefferson Abroad*. New York: Random House, Inc., 1999.

Photography and Illustration Notes

xviii author
3 author
4 canoe drawings (author); (bottom) University of Washington Manuscripts and Special Collections
5 author
6 (top left and lower right) Oregon Historical Society; (second from top) Columbia River Maritime Museum; (bottom left) Clatsop County Historical Society
9 Michael Haynes
13 University of Washington Manuscripts and Special Collections
16 (top) author, digital enhancements, Kate Hawley; (bottom) Columbia River Maritime Museum
18 author, digital enhancements, Kate Hawley
20 author
23 author
24 author
26 author
27 artifacts from Guest/McGlinn Collection (author's photo)
29 author
30 Oregon Historical Society
32 (top two images) University of Washington, Manuscripts and Special Collections; (bottom; right) Seattle Museum of Science and Industry
35 author
37 author
38 author
42 author
45 author
47 Artifacts from Fort Clatsop National Monument collections (author's photos)
50 author
53 author
58 author
59 author
62 author
69 American Philosophical Society
70 author
74 Clip Art
76 (top) author; (map) American Philosophical Society
78 Oregon Historical Society, 2 negatives combined, digital enhancements, Kate Hawley
82 author
83 author
84 Oregon Historical Society
85 Oregon Historical Society
86 Oregon Historical Society
87 (left) Clatsop County Historical Society; (right) University of Washington Manuscripts and Special Collections
92 artifacts from Guest/McGlinn Collection (author's photo)
93 artifacts from Guest/McGlinn Collection (author's photo)
94 author
97 author
98 artifacts from Guest/McGlinn Collections (author's photo)
101 author
103 Sonja May
105 Artifacts from Fort Clatsop National Monument collections (author's photos)
106 (top and right) University of Washington, Manuscripts and Special Collections; (bottom left) Clip Art
107 author
108 author
111 author
112 author
114 author
115 (both images) Chuck and Grace Bartlett
120 Chuck and Grace Bartlett
124 author
127 author
128-129 author
131 Guest/McGlinn Collection
134 author
137 Chuck Bartlett
138 Artifact from Fort Clatsop National Monument collections (author's photo)
140 author
151 "Stella" author
153 author
154 author
155 author
157 author, digital enhancements, Kate Hawley
158 author
161 author
163 author
165 Chuck and Grace Bartlett
167 Chuck and Grace Bartlett
168-169 Chuck and Grace Bartlett
173 author
176 author
179 author
183 author
188 N0AA chart of Columbia River
189 (top) Columbia River Maritime Museum; (bottom) Oregon Historical Society
190 (all maps) Oregon Historical Society
193 University of Washington Manuscripts and Special Collections
195 author (School ship *Danmark*, mid-Atlantic Ocean)
203 University of Washington, Manuscripts and Special Collections
204 University of Washington, Manuscripts and Special Collections
205 Chuck and Grace Bartlett

Journals

2, 8, 14, 22, 28, 34, 40, 44, 52, 60, 64, 72, 80, 88, 96, 100, 102, 110, 122, 126, 132, 136, 142, 146, 152, 160, 166, 170, 172, 178, 180 Reproductions of the authentic journals of Lewis and Clark used with permission from the American Philosophical Society.

Maps

Maps from several sources illustate the text:

69, 76 American Philosophical Society

188 National Oceanographic Atmospheric Administration

190 Oregon Historic Society

xii, xiii, xvi, xvii, 12, 13, 27, 63, 91, 92, 114, 184, 192, 203 author, colorization, Kate Hawley

All other maps derived from British Navigation Chart made under the Superintendence of Captain F.L. Evans, R.N., hydrographer, 1876. Colorization by Kate Hawley

INDEX

Italics indicate names on maps.

This book was written with Pilot Precise V-7 Rolling Ball pens.
It was designed on a Macintosh G3 Powerbook.
The fonts are Caslon, Minion Condensed, and Monet.